I0114717

'A beautiful, analytical and deeply human story of why reconnecting to the living biosphere is necessary, urgent, and provides us with the only path to resilience and harmony for all humans as an integral part of nature on planet Earth.'

—Professor Johan Rockström, Director of the Potsdam Institute for Climate Impact Research

'This is a terrific book on a profound question: how does the Nature around us influence the Nature within us? I don't know of another work that studies the question as deeply, nor with such elegance.'

—Professor Sir Partha Dasgupta, economist and author of *On Natural Capital*

'A brilliant, nearly poetic, call for a vital human reconnection with nature to save us all.'

—Thom Hartmann, author of *The Last Hours of Ancient Sunlight*

NATURE
WITHIN

James Bashford completed his medical degree at the University of Cambridge in 2010, before starting specialist neurology training at King's College Hospital, London. He obtained a PhD in clinical neuroscience in 2019. He now works as a clinical academic and honorary consultant neurologist in South London, leading a research programme and caring for patients in a specialist motor nerve clinic. *Nature Within* is James' first book.

Ways to follow James and keep in touch:

Bluesky: @wrenandbee.bsky.social
Instagram: @wrenandbee.nature
X: @wrenandbee
LinkedIn: james-bashford-neurology

www.wrenandbee.co.uk
www.healthconservation.org

linktr.ee/drjamesbashford

NATURE WITHIN

How the Natural World Shapes Our Minds, Bodies & Health

JAMES BASHFORD

PELAGIC PUBLISHING

First published in 2026 by
Pelagic Publishing
20–22 Wenlock Road
London N1 7GU

www.pelagicpublishing.com

Nature Within: How the Natural World Shapes Our Minds, Bodies & Health

Copyright © text, figures and images 2026 James Bashford

The moral rights of the author have been asserted by him in accordance with the Copyright, Designs and Patents Act 1988.

All rights reserved. Apart from short excerpts for use in research or for reviews, no part of this document may be printed or reproduced, stored in a retrieval system, or transmitted in any form or by any means, electronic, mechanical, photocopying, recording or otherwise, now known or hereafter invented, without prior permission from the publisher.

https://doi.org/10.53061/JGOL3050

A CIP record for this book is available from the British Library

ISBN 978-1-78427-593-8 Hardback
ISBN 978-1-78427-594-5 ePub
ISBN 978-1-78427-595-2 PDF

EU Authorised Representative: Easy Access System Europe – Mustamäe tee 50, 10621 Tallinn, Estonia, gpsr.requests@easproject.com

Cover image © George Peters/iStockphoto

Typeset in Minion Pro by S4Carlisle Publishing Services, Chennai, India

5 4 3 2 1

Contents

Peacock

For all the wrens and bees of the world

'We do not inherit the Earth from our parents,
we borrow it from our children.'

Indigenous American proverb

Acknowledgements

This book owes its existence to the help of many people. I want to thank all those who were sent versions of the manuscript in various states of disarray. Many provided insightful feedback that informed subsequent revisions, including Suzi Bashford, Adrian Broughton, Mel Cadman, Geoff Cunningham, Shane Delamont, Wyn Jones, Tom Killingbeck, Rob Murray, Lyn Ridgway, Mark Spurgeon and Greg Thompson.

Special thanks go to those named within the book, who have shared their expertise and enthusiasm with me directly. These include Tristan Bantock, David Cousins, Partha Dasgupta, Sam Lee, Geoff Martin, Miles Richardson and Sarah Walker. Any errors that remain are all mine.

I would probably still be mulling over my latest rejection if Nigel Massen and his excellent team at Pelagic Publishing had not taken on this project. For this great leap of faith and your passionate dedication to the natural world, I am immensely grateful.

I acknowledge the countless research and clinical colleagues I have had the pleasure of working with over the years. In particular, I owe a huge debt of gratitude to my supervisor and mentor, Chris Shaw, who put his trust in me over a decade ago as I set out on an academic career in neurology. And to all the funders of my research, thank you.

With the warmest of hearts, I thank my mum and dad, Lyn and Roy, who have always believed in me. And I thank my wife, Suzi, without whom this book would have been an impossibility. As we juggled young family life with busy work schedules, she allowed me to squirrel myself away from time to time to write the odd paragraph or two. Then, there are our wonderful twin daughters. Over the last eight years, they (more than anyone) have helped me recognise what the truly valuable things in life are.

Finally, I thank every patient who has ever shared details about their health in the hope I can make them better. Undeniably, this is what makes being a doctor one of our society's greatest honours.

Introduction

In our neglect of the natural world, we inflict the greatest self-harm

The Common Alder *Alnus glutinosa* is a pioneer of the temperate North. To achieve this, it has had to adapt. Since an individual tree produces both male and female flowers, the species is able to maintain its reproductive success in the face of its own population sparsity. And, unlike most other deciduous trees, its fruit is a cone, which disperses its buoyant seeds to the winds and nearby waterways, hence reducing its reliance on avian aid.[1]

Its real strength, however, is in the special relationship it has forged with the bacterium *Frankia alni*. This organism forms nodules in the tree's roots and specialises in fixing nitrogen gas directly from the air. Fixed nitrogen is then converted into ammonia, a redeemable currency for arboreal protein manufacture. In exchange, the bacterium receives sugars produced in the tree's leaves through the sun-driven, carbon-fixing process of photosynthesis. This reciprocal arrangement allows Common Alder trees to grow where others fail, enriching impoverished soils and initiating the natural architecture for life to flourish. Wherever Common Alders take root, other tree species, invertebrates, birds and mammals are sure to follow.[2]

The Common Alder–*Frankia alni* cooperation, or symbiosis, is just one elegant example of Nature's intricacy. These previously unconnected symbionts hit upon a way to enhance each other's survival, not with any foresight or agenda, but through an epoch-consuming process of trial and error. How many errors were needed before a successful formula was achieved? How many competing trials were wiped out by random environmental perturbations? Will this relationship one day cease to be an effective one owing to changing survival pressures?

Retrospective bias urges us to assume that Earth's living beings were destined to turn out the way they are today. But this is not the case. And this is what makes the natural world so special. Every kind of living being exists for the simple reason that it survives long enough to bear fertile offspring. If everything alive today had not proven this time and time again, it would not be here.

It follows that every living being is the carefully tailored result of its unique environment, interactions and exposures. By extension, there is no way to recreate what is lost. Why did large land-dwelling dinosaurs not re-evolve after their extinction 66 million years ago? This was not because dinosaur DNA was wiped out. We know this, because birds are modern-day descendants of the same family of dinosaurs that included the formidable *Tyrannosaurus rex*.[3]

Quite simply, dinosaurs did not re-evolve as before because the world was dramatically altered by the asteroid impact that led to their demise. And it just so happened that mammals adapted to the new terrestrial surroundings more effectively.

But as intriguing as interactions such as the Common Alder–*Frankia alni* symbiosis might be, specific examples like this represent only a speck on top of an iceberg's tip. And we can never hope to catalogue all of Nature's mysteries, as its myriad guises continually hone their survival-promoting attributes according to their environment.

What it means to be human

Homo sapiens, our species, is just one more item on life's conveyor belt of biodiversity. We are embedded within the natural processes that govern every other living being on this planet. It is just not possible to have any exceptions. And in a similar way to the Common Alder and its bacterial partner, our own evolutionary path has also relied upon a particularly nifty symbiotic invention: the mitochondrion.[4]

Billions of years ago, there existed the cellular prototypes of what would ultimately become a group of organisms known as the eukaryotes. Today, this group includes the animals, plants and fungi, with which we are all familiar. However, these early cells had a problem. They had limited means for internal energy production, thereby curbing their scalability.

But coexisting with these ancient eukaryotes were certain types of bacteria that had adapted the ability to make the ultimate of biofuels, adenosine triphosphate (ATP). If only there were some way of sharing the goods.[5]

The endosymbiotic theory posits that early eukaryotic cells engulfed these ATP-producing bacterial powerhouses, thereby conferring a substantial and long-lasting survival advantage. Over time, these engulfed bacteria became an intrinsic part of the eukaryotic cell. They became our mitochondria.

Evidence for this theory stems from the presence of a separate set of prehistoric DNA within each mitochondrion, which indicates its bacterial origin. Ultimately, these early eukaryotic cells had acquired their own energy security. This was quite the pioneering achievement.[6]

If you add this to the fact that the 30 trillion cells in each human body are matched in quantity by human gut bacteria, you start to wonder what proportion of *Homo sapiens* is truly 'human'. Even 8% of our genetic code today is composed of various extinct viral units that inserted themselves over the course of the last 100 million years.[7]

This serves to highlight one fundamental concept: humans are not, *and can never be*, external to Nature.

In fact, considering the origin of the mitochondrion, it is difficult to imagine a more integrated creature. We may have adopted ways to exploit Nature through the domestication of wild animals, the farming of monoculture crops and the killing of invertebrate 'pests', but this does not make us separate from it.

Nor do these exploitative acts make us unique.

Learning from the non-human world

Native to New Zealand, the Vegetable Caterpillar fungus *Ophiocordyceps robertsii* has evolved the lethal art of mummification. First, it releases spores that are ingested by porina moth caterpillars. When the porina caterpillar is ready to pupate and begin its transformation into a moth, the fungus encases it and consumes the contents. It then extends a fungal body from this lepidopteran tomb to release more spores, thereby propagating the deathly cycle.[8]

The trouble with our instinctively anthropocentric viewpoint is that the fungal incarceration of a caterpillar somehow seems different to the caging of chickens. But is it really? The method might be different but the subjugation of one product of Nature by another is just the same.

In fact, the progression of human exploitative behaviour is little more than copycat. Approximately 55–60 million years ago, just as the very first primates were evolving, leafcutter ants learnt how to farm. Their chosen harvest was a fungus. And as the ants foraged leaves to fertilise their new fungal gardens, the fungus proffered specialised hyphal endings packed full of sugars and fats, which were perfect for nourishing voracious ant larvae.[9]

But as impressive as this was, the ants did not have it all their own way. They were forced to purge their gardens of a fungal competitor that threatened to overthrow their dominion. And to do this, they required help from a bacterial partner that grew on their cuticles and produced an anti-fungal agent against the pests. But in response this unwelcome fungus learnt how to fight back with its own chemical arsenal against the ants and bacteria.[10]

Despite this evolutionary tug of war playing out on an almost incomprehensible timescale, a sense of harmony prevailed. Can humans hope to reach a similar state of equilibrium in our own farming and pest-killing endeavours? Yes, but only if we acknowledge that the rules of Nature cannot be outwitted.

Let's reconsider the expert of mummification once again. If the Vegetable Caterpillar fungus from New Zealand suddenly produced sufficient spores to infect an unusually large proportion of the porina moth caterpillar population, its reproductive success would be legion. However, the population of mating adult porina moths would be devastated, leading to precipitous declines in the numbers of porina moth ova and larvae. Consequently, despite the fungus' impressive release of spores, there would be no new caterpillars to consume them, and the spores would get blown or washed away. As a result, the porina moth population would be able to recover, restoring the delicate equilibrium between the two species.

In the most extreme scenario, the porina moth population may be too isolated and insufficiently robust to bounce back, leading to local extinctions. In turn, the fungus' initial boom would turn to eventual bust, underlining the perils of unsustainable living, irrespective of the ecosystem niche you occupy.

This is a sobering thought when one considers that the current rate of human activity would require 1.8 Earths to continue sustainably. Earth Overshoot Day is the day when the ecological regenerative capacity of our Earth is exhausted for that year. In 2025, this fell on 24 July.[11]

What this fungus–moth interaction emphasises is that there is a series of natural checks among interdependent organisms to maintain healthy populations of each. Scale this up to the planetary ecosystem level and the number of codependent interactions stretches far beyond current scientific knowledge. Piecemeal, the sustainability of the system is largely maintained, threatened only when stressors on it mount at a disproportionate rate.

The natural rhythms of life

The parched elephant in the sultry room is global warming. I am not referring to the historic periods of global scorching that the Earth has experienced at various times in its 4.6-billion-year life, the last of which occurred 55 million years ago and produced average global temperatures over 12 degrees Celsius warmer than today. I am instead referring to anthropogenic global warming driven by the carbon catapult of modern life.[12]

While the previous rises in global temperature typically crawled over many millions of years, the one we face today has led to a 1.2 degrees Celsius rise in just 150 years. This is putting unspeakable stress on biological systems over a time frame that is much less conducive to natural adaptation.

Whereas the slow drift of survival pressures on leafcutter ants tens of millions of years ago afforded them the time to galvanise symbiotic recruits over countless generations, the survival challenge posed by anthropogenic global warming and its associated activities today is advancing at an unprecedented rate. Sadly, for recently extinct species such as Australia's Paradise Parrot *Psephotellus pulcherrimus* and Costa Rica's Golden Toad *Incilius periglenes*, it has advanced at an irrecoverable pace. Even more worryingly, as the biodiversity of an ecosystem shrinks, so does its resilience to further incursions. And a vicious cycle of accelerated extinction supervenes.

In *A Short History of Nearly Everything*, Bill Bryson famously condensed the whole of Earth's history into a single day. Despite the emergence of single-celled life at about 4:00:00 a.m. on that day, dinosaurs do not appear until just before 11:00:00 p.m., some 19 hours later, before suddenly vanishing at 11:39:00 p.m. The first human evolves at 11:58:43 p.m., just one minute and seventeen seconds before midnight, with the whole of our recorded history lasting 'no more than a few seconds'.[13]

When considered on this more relatable scale, it becomes clear that our instinctive concept of time is incredibly skewed. It evolved this way as there was no intrinsic survival benefit to understand the gargantuan linear timescales involved. Instead, our brains and physiology are much better attuned to cyclical patterns: the daily spin of the Earth on its axis; the near-monthly orbit of the Moon around our planet; and the annual revolution of the Earth around the Sun. These cycles set the rhythms of all known life.

Time for some self-reflection

If one accepts that at the fundamental biological level there is nothing that distinguishes humans from any of Nature's progeny, regulated by the same natural laws, then this has significant ramifications for human health. Our physiology, biochemistry and anatomy, right down to the subcellular domain, have been shaped and sustained by natural interactions for thousands of millennia. And so any form of disconnection, particularly a relatively rapid one, is bound to have unintended and distasteful consequences.

The urbanisation of society, the digitalisation of recreation, the relentless pursuit of economic growth and the destruction of natural habitats are all contributors to the modern human detachment. And our technological success has promoted a hubris that is difficult to shrug off. Our global society acts as though the best thing for us is some sort of 'naturectomy', driving us further and further from our roots.

This book is about exploring the multifarious reasons why this would be the very worst outcome, not only for the health of our own species, but also for the health of all the precious elements of Earthly life.

In Part I, I introduce the core issues at play, beginning with an unexpected woodland encounter that made me think more deeply about how we sense the natural world (Chapter 1). In Chapters 2 and 3, I outline the fundamental roles that knowledge and our perception of disease play in any discussion about health. Then, in Part II (Chapters 4 and 5), I explore some key features of our planet's workings and how life interacts with the only place it can call home. For Part III, I home in on the value of biodiversity (Chapter 6) and the kindness of Nature (Chapter 7). Finally, Part IV looks ahead. In Chapter 8, I consider how we might avert the full effects of a less biodiverse, less natural world; and Chapter 9 draws on the importance of a much-needed reset.

Ultimately, this book is about a species in desperate need of some self-reflection.

PART I

MAKING SENSE OF WHAT WE KNOW AND HOW WE FEEL ABOUT THE NATURAL WORLD

Lappet

CHAPTER 1

The Veery's Duet

In both ears and out of nowhere

The flutey, tumbling rounds of *vee-eer-eer-eer-eer* whistled brusquely yet breezily through the leaves. Although the immediate harshness pierced the prevailing quietness like a sword through jelly, each round promptly mellowed and earned its right to command the auditory void. While the message had an alien quality, it seemed as if this messenger came in peace. Part distress call, part hypnotic lilt, the auditory juxtapositions of this impromptu song were stark.

It was Saturday 25 June 2022, and I was in the Réserve Naturelle du Marais-Léon-Provancher in Québec, Canada. I had been ambling around the reserve for a couple of hours, engaging in a leisurely search for species of birds, butterflies and day-flying moths yet unencountered by my neural networks. The early afternoon was hot and humid enough that even short spells in the sun were undesirable, and so I was pleased when the uncovered path meandered off into unspoilt woodland perched above the St Lawrence river and its lowlands.

It was soon after entering this part of the reserve when I heard the song that stopped me in my tracks. I listened closely, trying to work out its source. I instinctively pointed my camera in its direction, flicked the video on and softly edged closer. As I followed the snaking path among the trees, keeping my eyes peeled, I was glad to hear the song becoming louder. With each step, I became increasingly eager to catch a glimpse of the woodland singer. Equally, I feared each next step might frighten it away.

Luckily, the singer tolerated my steady approach, and I finally set eyes on the bird producing the unique song. At that moment, I had no idea of its identity, only later discovering its name. Not only was this the first time I had ever heard a Veery *Catharus fuscescens* sing, I confess this was the first time I even knew this species of bird existed. And the resounding influence of that anonymous encounter has since become a symbol of my own reconnection with the natural world.

My reason for being in Canada was work related. As a clinical researcher in the field of neuroscience, I had been presenting recent results at an international congress in Québec City. After a heavily delayed flight from the UK, the week had been much more tiring than anticipated. I felt mentally fatigued, and my attentive capacity was in desperate need of a recharge. And so, on the final day of my trip, I headed out of the city to the *réserve naturelle*. But why should my instinct to seek out Nature make me feel better? There are at least three theories to help explain.[14]

The attention restoration theory posits that exposure to Nature induces something called quiet fascination. This amounts to the satisfaction you experience when you passively let the world pass you by. Instead of directing your attention to a narrow set of tasks, constantly inhibiting the numerous distractions of modern life, quiet fascination allows your brain to expend much less cognitive energy. This gives your brain time to refresh its attentional networks, thereby staving off mental fatigue.

Therefore, as I strolled around the reserve, letting the natural environment pass me by, I was benefiting with each step. I met the hauntingly named Toothed Somberwing moth *Euclidia cuspidea*, the doubly named Camberwell Beauty/Mourning Cloak butterfly *Nymphalis antiopa* and the impressive Pileated Woodpecker *Dryocopus pileatus*. My quiet fascination with Nature was in full flow.

The second theory to mention is the stress reduction theory. This states that exposure to Nature brings about a prompt alleviation of the acute stress response, resulting in normalised blood pressure and reduced feelings of anxiety. This is at the heart of the desire to get some fresh air in the wilderness at the end of a stressful day.[15]

When our innate stress response is activated by modern aspects of living without reprieve, the cumulative burden on our physiological and cognitive function can become disabling. It is like having a national terror threat level continuously set on high alert, utilising scarce energy, time and resources inappropriately and unsustainably. Interactions with Nature allow you to reset your inner threat level to a more relaxed state.

The third theory concerns an instinctive preference for landscapes that offer both prospect and refuge. It is known as the prospect–refuge theory and can be thought of like this. As I entered the wooded area of the reserve, the vista over the lowlands permitted surveillance of the wider environment (the prospect). Such an outlook would have given my prehistoric ancestors a survival advantage, as they scanned for both prey and predator. Meanwhile, the woodland offered my in-built survival networks the promise of imminent seclusion should it be needed (the refuge).[16]

This theory reminds me how dependent our brains are on the genetic legacy of our evolutionary past, a concept that motivates me to better understand the interconnectedness between human health and the natural world.

But while I valued these general theories about our connection with the natural world, I felt they only touched the surface of what was really going on. To learn more, I first needed to reflect on how we take in our surroundings.

Sensing the world around us

There are five main access routes for external stimuli to reach the mass of 170 billion cells that make up the human brain. These are the five senses that babies are busy honing from birth: vision, sound, smell, taste and touch. So ingrained is our ostensibly complete understanding of these fundamental biological mechanisms that having a 'sixth sense' has been adopted as an unexplained, often intuitive, ability for worldly perception. Whether this sixth sense confers the power to see dead people or simply to feel you are being surreptitiously watched as you walk along the street, I would contend that it is the five main senses that are responsible, buoyed by the emotional context, some misperceived coincidence and a certain 1999 Hollywood blockbuster.

Let me introduce these five senses in turn, starting with vision.

Relying heavily on visual inputs for our interaction with the outside world, the human brain devotes significant space and energy to the visual processing network. Light travelling at 300,000 km/s enters the eye through the pupil and activates photoreceptors in the retina. From each eye, the second cranial nerve conveys retinal activity to midbrain relay centres, before the information pathway radiates onwards to the primary visual cortex at the back of the brain. Further connections to higher order visual processing cortices impart complex details regarding the what and the where of the scene around you.[17]

Primed for almost constant activity during waking hours, the visual system is capable of producing fake images known as hallucinations. An interesting example is a condition called Charles Bonnet syndrome, named after an eighteenth-century Genevan naturalist. In this condition, complex hallucinations result from a deprivation of visual stimuli in individuals with poor eyesight. Starved of their usual electrical inputs from the eyes, the intricately evolved neural networks find it difficult to remain quiescent, prompting one to question what proportion of what we 'see' can be entirely relied upon.[18]

Unfortunately, there are a multitude of neurological conditions that can disrupt the visual pathway, including stroke, multiple sclerosis and migraine, all of which are common reasons for referral in my practice as a neurologist. And testament to the high energy burden of the visual system, it is essential to exclude new stimuli every night during sleep, allowing a period for consolidation of the day's events and neuronal maintenance.[19]

In contrast, the portal for sound, the ear, remains permanently open, and therefore hearing has become a great survival asset during sleep. Who hasn't been woken up at 4 a.m. by a loud bang, the taunting screech of a fox or a nocturnally playful cat with one's heart pounding in an immediate state of bewildered alertness? Crucially, the ear surveys the parts of the surrounding environment inaccessible to the eye.

The auditory pathway is just as specialised for its role as its visual counterpart. Sound waves are funnelled by the ear canal towards the ear drum, which sets off an amplifying mechanical relay in the three smallest bones in the human body: the ossicles. As energy flows from hammer to anvil to stirrup, ripples along a spiral of fluid are set up within the cochlea (meaning 'snail shell' in Latin). These ripples are then detected by small hairs and converted into electrical signals, which travel along the eighth cranial nerve to brainstem centres before progressing to the auditory processing centres on both sides of the brain. Subtle time differences between signals arriving from each ear permit the localisation of sound.[20]

This system can be duped when visual information tells a different story, though. Take the art of ventriloquism as an excellent example.

The mechanics of sound waves travelling through the air at 330 m per second do not change, but when the performer misdirects your attention through the combination of an animated puppet and lipless vocalisations, the visual trumps the audible.

The yin and yang of sound processing leads on one hand to noise pollution, noise-induced hearing loss and tinnitus, but on the other to relaxation, enjoyment and meditative reflection. While the positive effects can be induced by sounds that are either natural (e.g., birdsong, lapping waves) or human-made (e.g., Mozart's sonata for two pianos in D major, K. 448), the negative effects are almost the exclusive product of non-natural auditory input.[21]

Even the absence of sound has an impact on our brain function. It turns out that our brains are primed to listen expectantly in the quietness. While it would be reasonable to presume that silence is always golden, the absence of natural sounds such as birdsong most likely served as a potent alert to our ancestors, who would have automatically linked this with a potential predator nearby.[22]

What is clear is that sounds in their various pitches and volumes (or lack thereof) can have a significant impact on our physical and emotional well-being. And this has huge importance when engaging with the natural world.

Next up is smell, more formally known as olfaction. And as the most primordial of our five senses, it boasts the most direct connection through to the brain. Airborne, volatile odours enter the nose and are detected by specialised olfactory sensory neurons packed together at a density of 3 million per cm^2. These project through small holes in the skull towards two antenna-like extensions of the brain known as the olfactory bulbs. Underlining smell's importance in human emotion, the primary olfactory processing centre is housed right next to our fear centre.

Whether it is the rotting flesh odour of putrescine or the metallic mushroomy aroma of 1-octen-3-one, each chemical activates a unique spatial pattern of olfactory sensory neurons, which is then recognised by the brain like a QR code. Some compounds such as indole and skatole are Machiavellian, in that they engender the sweet smell of orange jasmine at low concentrations, but at high concentrations invoke the pungency of mammalian faeces.[23]

Conserved across mammals as a survival-enhancing tool, it might seem that smell is not as useful to modern humans as it would have been to our ancestors. And in the context of aiding the identification of offensive and potentially harmful food before ingestion, this might be true. However, the role of smell extends far beyond this, influencing our mood, attention, memory, situational assessment, appetite, libido and social interaction, largely at the subconscious level. Pheromones, once thought erroneously to play only a minimal role in humans, act on an accessory pathway that complements smell, providing a powerful means for communication between individuals of the same species.

Meanwhile, taste, or gustation, is often considered smell's ill-equipped understudy, rendering even the most fulsome dishes unpalatable while olfaction is temporarily out of action. But the truth is thankfully much more savoury than that. Taste buds are exceptionally good at what they do: tasting chemicals. It's just that this chemical taste is only a part of the overall flavour you perceive when tucking into your favourite meal. The smell of food is a significant factor, but then so are the visual, the auditory and the tactile. Therefore, like any interaction with the outside world, the brain assimilates inputs from the full sensorium to produce an enriched perception of the food we taste.[24]

With varying thresholds for sour, salt, sweet, bitter and umami, the adult human tongue plays host to 6,000 taste buds on average. Each one is a specialised group of cells capable of converting chemistry into electricity, sending its message via dedicated taste nerves to the brain. While imaginary tastes classically occur during the aura before an epileptic seizure, this should not be considered the exclusive remit of gustation. Theoretically, anything the brain can do, a seizure can do too, only more intensely and uncontrollably.

The final sense to introduce is touch. When the largest organ in your body, the skin, comes into physical contact with an external object or being, several tactile properties are sensed and conveyed to the central nervous system. These include pressure, vibration, temperature and pain, and all these modalities comprise the sensation of touch. I have used the broader term 'central nervous system' purposefully here, as not all these signals make their way to the brain. Instinctively, you pull your hand away from a sharp object to avoid injury. As brain involvement would waste unaffordable milliseconds, this survival-enhancing ability has evolved as a simple reflex arc that only gets as far as the spinal cord.[25]

Nonetheless, most tactile signals do reach the brain, assimilated within complex processing networks. Sometimes, simple observations emerge above this complexity. One such example is a curious phenomenon known as a sensory trick. This is where individuals with usually uncontrollable spasms in the neck can dramatically improve their symptoms by lightly placing a hand on the affected area. How this occurs is not well understood, but it emphasises the powerful effects of all sensory inputs on brain function.

With all this in mind, my thoughts now return to the benefits of an attention-restoring, stress-reducing stroll through a Canadian woodland.

Nature keeps us grounded

Reliant on my five senses, my exploration of the *réserve naturelle* yielded many more species besides the Veery. These included the tongue-twistingly termed Greyish Zanclognatha moth *Zanclognatha pedipilalis*, the mosaic-patterned Pearl Crescent butterfly *Phyciodes tharos* and the elegant Tree Swallow *Tachycineta bicolor*. Although it is species such as these that steal the limelight and consume my camera's memory, I am increasingly aware of the multitude of intervening moments between each click of the shutter. After all, the camera brings a modern artifice to what ultimately is an ancient exercise.

While a photograph is the return ticket to a moment passed, what if the capture of that time-travelling permit alters the natural storage mechanisms at play? Maybe a lack of full engagement in the moment undermines the original purpose of immersing oneself in Nature. This is akin to the observer effect in physics, which states that the very act of measuring something perturbs the measurement itself.

Of course, focusing on the second-to-second present is the underlying principle in mindfulness techniques. And Nature offers an amazing arena for inducing a mindful mind. There is just so much to zone in on: the crisp crunch of autumn leaves underfoot, the delicate caress of a warm breeze on the face, the Caravaggesque dance of light and shadow on the woodland floor, the pleasing aroma of petrichor after rainfall, the aquamarine blur of a hunting kingfisher and the repetitive riff of a returning chiffchaff in spring. Novelty stimulates a sense of fascination in what is unfathomably complex, while familiarity invokes a cumulative comfort that all is stable in the world.

I am not suggesting that every stroll through the natural environment is a poetic ideal of pleasant and novel stimuli. In fact, it is essential to experience a full spectrum of exposures to apply a meaningful range of reference. In a modern world crammed full of stimuli arriving at our brains over increasingly shorter timescales, it is easy for our minds to become detrimentally skewed. As a result, expectations about the future are inflated, and when reality does not fulfil those expectations, negative emotions such as disappointment, sadness and despair ensue.

Let's consider the doughnut, or any human-devised vassal for concentrated sugar delivery. The doughnut has only existed for a negligible portion of human evolution, and yet it has been ingeniously designed to spark an ancient emotional network that was seldom, if ever, stimulated to such a degree.

At the biochemical level, the sweet taste of a doughnut provides an instantaneous dopamine hit to areas of the brain that confer what psychobiologists call reward. While our ancestors would have been forced to work hard for their equivalent, naturally sourced reward (say, a morsel of honey), the modern human in the developed world does not have such an arduous task.[26]

For our predecessors, the high calorific content of sugar would have been invariably paired with an energetic and lengthy foraging or hunting exercise, thereby replenishing depleted carbohydrate and fat stores. For the twenty-first-century human, it is possible to obtain the reward by doing as little as lifting one's arm and opening one's mouth. This disparity between the task and reward means the brain's emotional and physiological networks can easily become unbalanced, either because the stimulus is too intense, too frequent or both. There are plentiful examples of such modern-day supranormal inputs across the full sensorium. Having either positive or negative effects, these range from the immersive cinema experience to doomscrolling on social media.

Nature has a grounding effect, because it shifts the perception of stimuli back to its normal range. 'Normal' is often difficult to define precisely, but in this context a pragmatic approach based on our evolutionary foundations is useful.

As the human brain is the perfectly tailored product of our ancestors' immersion in the natural world over millions of years, it stands to reason that it copes best when exposed to the same assortment of experiences that guided its development. When operating within this 'normal' natural environment, the brain can successfully rely on a set of assumptions about how the world works. Unfavourable fluctuations within the confines of these natural laws are typically well managed, bringing about resilience and fortitude. Equally, while more desirable variations are avidly sought, their occurrence is not immediately forthcoming, resulting in tenacity and well-earned reward. It is therefore little surprise that this same set of assumptions performs imperfectly when adopted in a more unnatural setting.

Autism spectrum disorder is a neurodevelopmental impairment that, among deficits in communication and social interaction, restricts the ability to assimilate sensory inputs. Consequently, individuals suffer from a range of stimulus hypersensitivities, leading to a more rigid way of living as an avoidance strategy. Stereotypical behaviours such as body rocking, hand flapping, blinking and repetition of sounds help to shield the individual from the source of hyperarousal. And solace is often sought through the predictability of Nature or focused activities that minimise external interruption.[27]

One could view this phenomenon as a relative lack of resilience to the supranormal stimuli that are commonplace in modern society. It is probably an extreme case of what most of us regularly experience to varying degrees. Recent evidence supports this idea, showing that the close association between stimulus hypersensitivity and repetitive shielding behaviours was found not only in children with autism spectrum disorder, but also in typically developing children, albeit to a lesser degree in the latter group.[28]

Therefore, it has been suggested that the effects of hyperarousal and hypersensitivity are also relevant to everyday stimuli among the wider population. Grounding activities, such as hobbies, socialising and exercise, are purposefully built into our lifestyles as a way to limit these deleterious effects. And when these activities overlap with Nature, the brain is grateful to return to its ancestral home.

The human brain is a black box

At this point, it is worth considering what happens within our brains in a bit more detail. The grey and white matter of the human brain can be modelled as a black box with defined inputs and outputs (Figure 1). The five ways of sensing external stimuli are processed in ever more

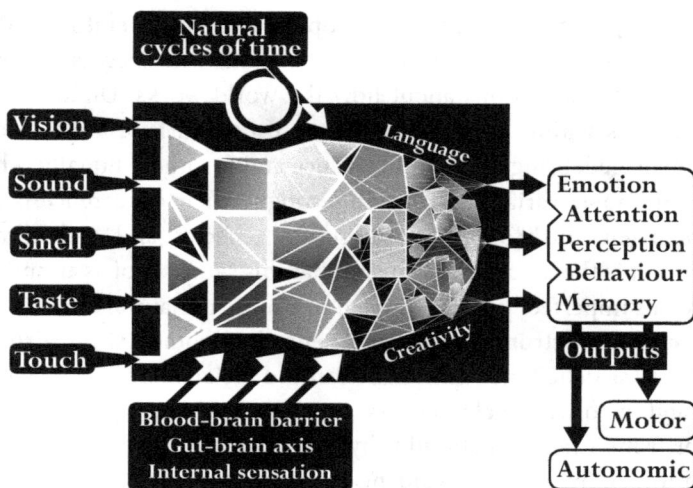

Figure 1. A black box model of human brain function.

complex ways as the number of interactions between them increases. These dynamic pathways have been forged by millions of years of genetic evolution and are constantly adapted by life experience and the information that flows along them. This dictates the relative proportions of neurotransmitters in specific functional compartments. For example, higher levels of dopamine and serotonin confer more positive aspects of brain function, such as reward and happiness.

It is a general principle that the neural patterns that are repeated get reinforced, while those that are neglected become weakened. Let's consider a tangible example. When you play tennis, your brain predicts the complex set of movements required to return the ball, obtaining immediate feedback on the success of each prediction it makes. The pattern of movements that led to a passing shot skimming off the baseline to win the set gets strengthened, while the pattern that sent the ball flying over the fence is repressed. Practice makes perfect, or at least a little less imperfect than last time.[29]

Alongside the five senses for processing external stimuli, there are also internal signals arriving at the brain via three main routes: the blood–brain barrier, the gut–brain axis and the internal sensory system. Let's briefly consider each of these.

First, the blood–brain barrier controls which molecules are allowed to permeate the neuronal milieu, including sugars, proteins and metals, providing essential components for normal cerebral function. Importantly, the array of constituents arriving at the brain in the blood is influenced by diet, hormone levels, state of health, emotion and pharmacological additives. Second, the gut–brain axis refers to the host of bacteria in each of our gastrointestinal tracts that is increasingly appreciated for its direct influence on the brain. Finally, the internal sensory system is a collection of inputs from the muscles, tendons, viscera and inner ear that act as the sentinels for bodily deviations, informing the brain about muscle stretch, joint position, gut distension, internal pain and head orientation. In particular, these inputs provide essential information for the coordination centre at the back of brain, known as the cerebellum (meaning 'little brain').[30]

Now turning our attention away from the brain's inputs, the outputs produced by this black box are what make us who we are and determine how we act and feel. They lie on a spectrum from more intrinsic outputs such as perception, memory and emotion to more extrinsic features such as attention and behaviour. They are mediated via two main streams of information flow, which in classical teaching are represented by the dominant language hemisphere and the non-dominant creativity hemisphere. Broadly, the dominant hemisphere is more associated with logical reasoning and planning, while the non-dominant hemisphere is more responsible for spatial awareness and imagination.[31]

And mediating these outputs to the outside world is the remit of two distinct parts of our nervous systems: the motor pathway and the autonomic nervous system.

The motor pathway receives guidance from the cerebellum before providing the output for voluntary control of our muscles. This final set of instructions from the brain is required each time you walk, talk, grip, skip, drink, blink, dance, prance, write, bite, sing, cling, cry or sigh.

Meanwhile, the autonomic nervous system is of critical importance when it comes to our behavioural expression to the outside world. Divided into the sympathetic fight-or-flight system and the parasympathetic rest-and-digest system, it is not only responsible for getting you to your imminently departing train just as the doors are shutting but also for the post-prandial drowsiness that plagues Sunday afternoons.[32]

Natural cycles of time entrain the brain

So far in this chapter, I have focused on the inputs and outputs of this black box model of the brain. But there is one set of influences on brain function that deserves special mention: the natural cycles of time.

Astronomical periodicity related to the relative positions of the Earth, Sun and Moon over their 4.6-billion-year history has had an inextricable bearing on the evolution of life. As the Earth spins on its axis, the ability to alternate between restful night-time phases and active daytime states according to external luminary cues is crucial for diurnal animals such as ourselves.

Present in the brains of all mammals, a master timekeeper tightly regulates the daily rhythm of life. In humans, this is mediated, in part, by the rousing stimulus of blood cortisol in the mornings and the sedative influence of melatonin in the evenings.[33]

It is important to note that our species spent most of its evolutionary development in African localities close to the equator, only sustainably emerging to more temperate zones from about 60,000 years ago. Therefore, the pronounced seasonal variations in daytime light exposure experienced by the majority of humans living today represent a relatively new environmental challenge.[34]

Indeed, for many individuals it is a physiological strain too great. Seasonal affective disorder manifests as a recurring depressive state peaking in autumn and winter months owing to a combination of reduced light exposure and prolonged melatonin secretion. This latter factor depletes the brain of one of its most potent happy hormones, serotonin, as this is what melatonin is made from.[35]

Furthermore, a seemingly unconnected condition of the nervous system, multiple sclerosis, is also associated with life away from the equator. In this disease, the immune system attacks the natural fatty coating of nerves that in health functions to speed up nerve transmission. Patients typically suffer from intermittent symptoms such as painful visual loss, balance difficulty, bladder control disturbance and limb weakness, but over time a progressive decline in neurological function often predominates.[36]

Several observations make it clear that the risk factors predisposing to multiple sclerosis start early in life, even before birth. For example, those born in the spring at higher latitudes are at increased risk, thought to be due to low sunlight exposure during a predominantly autumn- and winter-based pregnancy. And even among those born near the equator, a move to higher latitudes before the age of 15 increases the risk of developing multiple sclerosis by up to 80 times.[37]

Clearly, there is something about exposing ourselves to a diurnal and seasonal pattern of sunlight distinct from the one experienced by our equatorial ancestors that can be deleterious to our health. The fact that seasonal affective disorder and multiple sclerosis exist at all indicates a misalignment between what the human brain has been used to and what modern life supplies it with. Perhaps further insight can be illuminated from the other key celestial object related to our planet: the Moon.

On 11 October 1492 at about 10 p.m., with the *Santa Maria* poised for its historic landfall at San Salvador Island in four hours' time, Christopher Columbus and his crew espied the 'flame of a small candle', incandescent in the marine darkness. The timing and location of this natural spectacle align with the captivating mating ritual of the Bermuda Fireworm *Odontosyllis enopla*, demonstrating one of the clearest examples of lunar periodicity in Nature. Peaking one hour after sunset on the second or third night after full moon during summer and autumn months, sexually mature male and female fireworms swarm together to release their sperm and eggs in a bioluminescent swirl of procreation.[38]

How marine worms achieve such exquisite lunar synchronisation is becoming clearer. In research on the related bristle worms, a light-receptive protein has been discovered that is instrumental in this process. This protein is able to distinguish moonlight's glow from sunlight's glare, suggesting predictable cycles in ambient light levels help to entrain the reproductive cycle of the marine worms.[39]

Despite this clear link between the Moon and marine worms, evidence linking human biology with the lunar cycle is less convincing. At first glance, it seems suspiciously coincidental that the human menstrual cycle approximates to the 29.5-day lunar phase cycle. Indeed, early studies in the 1950s investigating hundreds of thousands of births indicated a marginal excess of births around the full moon and a corresponding dip on moonless nights. However, many modern factors such as the urban ubiquity of artificial light, the global variability in cloud cover and the influence of maternity care make research in this area challenging. In fact, subsequent data have *not* consistently supported a link between birth rate and the lunar cycle.[40]

Although the influence of the Moon on the human reproductive cycle remains unclear, a credible connection with our sleep patterns is now emerging. More pronounced in males, the days just before the full moon are associated with reduced sleep duration and delayed sleep onset, corresponding with lower night-time melatonin levels and higher cortisol release. This fits the hypothesis that those ancestors of ours who best capitalised on moonlit nights for predator vigilance and prey capture would have enhanced their survival. However, the discovery of a human gene analogous to the lunar clock found in marine worms remains elusive.[41]

For me, the most striking thing about the black box model of brain processing is its unknowable complexity. But then Nature has been working on it for millions of years in some form or another, so perhaps this should come as no surprise. And when faced with a cipher whose inner calculations are enigmatic, it seems prudent to focus on what goes in and what comes out.

From a core biological standpoint, our brains have been essentially unaltered by the technological advancements and urban expansion that have dictated the last few millennia of human history. Modern humans possess the same five methods for sensing the outside world, influenced by the same internal mechanisms and the same natural cycles of time as our pre-modern ancestors. The pragmatic fallout is that if we desire optimal outputs from our brains today, then an improved appreciation for their naturally sustaining inputs is vital.

The making of a song

Let me now return to the woodland experience I opened this chapter with. But instead of exploring the ways our brains sense the natural world, I want to focus on the song itself.

On first hearing the ethereal *vee-eer-eer-eer-eer* of the Veery, I stood transfixed. Immediately, my memory scrambled for any records of such a sound. Its answer was a surprise to me: the whirly tube. As a plastic toy I remembered from my childhood, it was about the least natural object imaginable. And yet that was what the melodic notes conjured up in my mind as I stood listening among the trees.

Growing into a New York craze in the 1970s, the whirly tube is a long cylinder of corrugated plastic that produces multiple pitches simultaneously when rotated through the air. The higher air pressure at the handheld end flows down the ridged tube, setting up a vortex of harmonic resonance.

So mesmerising is the timbre of this simple instrument, an ensemble of whirlies has been described as 'sometimes hauntingly beautiful, sometimes dramatic, sometimes soft, sometimes strong and robust, but at all times inspiring and thought provoking'. This is perfectly portrayed by Matt Miller's 2018 original score entitled *Music for Whirly Tubes*. Once heard, it is easy to understand why this instrument boasts the endearing epithet 'lasso d'amour'.[42]

In June 2021, the G7 leaders met in Cornwall, England. On this temporary centre stage of global politics, a Cornish beach witnessed a 1,000-strong orchestra of whirlies hypnotising onlookers with *Kan an Mor (Song of the Sea)*. Using whirlies recycled from used fishing nets, this sustainable metaphor highlighted to world leaders the vital and overdue need for ocean detoxification. With this in mind, the whirly tube was maybe not as unnatural as I originally thought.

Nevertheless, as I have already explained, the song I heard in the Canadian woodland was sung by a bird known as the Veery, not by a whirly tube. But to explain their auditory resemblance, it is necessary to delve into the specialist anatomy of the avian respiratory system.

Birds have evolved a unique structure for vocalisation called the syrinx. Whereas mammals, reptiles and amphibians rely on the closely related larynx for voice production, the avian larynx only exists today as an inaudible relic. For birds, the switch from larynx to syrinx resulted from a strong evolutionary pressure for song diversification.[43]

Customised to the survival demands of each species, a multitude of vocalisations occur among the bird population: to raise the alarm to predatory danger; to demarcate territory between competitors; to signify edible bounty to kin; and to lure mates during courting rituals. These methods of communication have been tailored for specific habitats, whether operating in the midst of the thickest forest or over the bare savannah.

The timing of birdsong is important, too. Birds are in tune with the natural cycles of time to such an extent that humans (and other animals) have learned their daily and seasonal patterns. In Europe, the spring is so renowned for its impressive morning performances from the likes of the Song Thrush *Turdus philomelos*, European Robin *Erithacus rubecula* and Eurasian Blackcap *Sylvia atricapilla* that International Dawn Chorus Day is held on the first Sunday of May each year. And the arrival of the onomatopoeically named Common Chiffchaff *Phylloscopus collybita* in the UK from their overwintering retreat in Africa ushers in the distinctive sound of spring.

In the short story 'Escape from Spiderhead', George Saunders perfectly summarises the important role birdsong plays:

> Night was falling. Birds were singing. Birds were, it occurred to me to say, enacting a frantic celebration of day's end. They were manifesting as the earth's bright-colored nerve endings, the sun's descent urging them into activity, filling them individually with life nectar, the life nectar then being passed into the world, out of each beak, in the form of that bird's distinctive song, which was, in turn, an accident of beak shape, throat shape, breast configuration, brain chemistry: some birds blessed in voice, others cursed; some squeaking, others rapturous.

So what has allowed birds to develop such status as Nature's chimes? While their switch from larynx to syrinx was vital, they have also relied on a breathing technique that sets birds apart from all other animals alive today.

Taking the deepest of breaths

When birds breathe, they utilise a continuous flow-through system for gas exchange. This is made possible by the presence of bellowing air sacs that keep the flow of air moving in only one direction through their lungs. This is in contrast to the back-and-forth tidal system found in mammals (Figure 2).[44]

One of the problems with the mammalian approach is that it penalises so-called dead space. This is the proportion of inhaled air that never participates in gas exchange, simply because it never

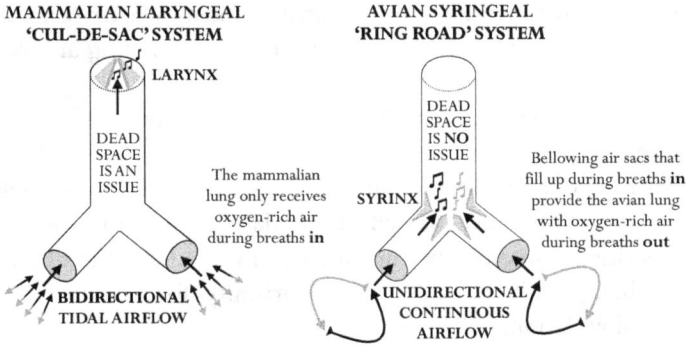

Figure 2. Sound production in the mammalian and avian upper respiratory tracts.

arrives at the lungs. This is important, as more dead space means more energy is wasted shifting air in and out of the respiratory tracts without any hope of absorbing its life-sustaining oxygen or depositing waste carbon dioxide into it. In humans, dead space equates to 30% of the total air we breathe, making it relatively inefficient.

And this inefficiency has put evolutionary constraints on the mammalian anatomical blueprint. For example, the only way a giraffe can reach untouched leaves with its long neck is by significantly reducing the girth of its extended windpipe. This adaptation has evolved to limit the inevitable increase in dead space associated with a longer windpipe. Although there was evidently a strong evolutionary pressure for an elongated neck in terms of foliage accessibility, one undesirable knock-on effect for the giraffe is that it restricts the range of sounds it can produce. In fact, its narrow windpipe causes such an increase in air resistance that it can barely generate the air pressures required to make sounds we can hear.

Luckily, giraffe evolution has been much more reliant on the perfectly adapted giraffe ear capable of hearing these hushed, deep tones than it has on its ability to communicate with humans. So if you are ever asked what sound a giraffe makes *and* you are a human, silence is the most appropriate response.

Despite this issue with dead space, the mammalian blueprint has certainly been capable of reaching what we would consider colossal sizes today, such as the recently discovered fossil of the extinct giant rhino, estimated to weigh the equivalent of six modern elephants. But this is still three times smaller than the largest known dinosaur, the titanosaur. Remember that the titanosaur would have had a respiratory system akin to modern birds. So let's get back to looking at how the avian lung achieves enhanced respiratory efficiency.[45]

Despite the relatively small size of the avian lungs, their success relies on the arrival of a constant stream of fresh air even when the bird breathes out. This is all thanks to those bellowing air sacs, which create a non-stop conveyor belt of circulating oxygen to each lung. And it's this adaptation that has provided birds (and the dinosaurs before them) with a wealth of opportunity for anatomical and functional variation.

Unlike the mammalian giraffe, it is no problem for a bird such as the White Stork *Ciconia ciconia* to have a long neck, as its method for respiration does not penalise a long and broad windpipe. In fact, the windpipes of birds are approximately 2.7 times longer and 1.3 times wider than those of comparably sized mammals. This unique anatomy was a permissive factor for the evolution of the gargantuan, long-necked dinosaur family known as the sauropods, while it also facilitated the energy-consuming task of flying in birds. Most important to the production of birdsong, this respiratory setup promoted the advantageous evolutionary switch from larynx to syrinx.[46]

Perfecting the harmonic duet

Before delving even further into the mechanics of birdsong, it is helpful to understand the basics of human voice production.

In the human larynx, a single pair of taut vocal cords are vibrated by the movement of air across them. As you push air out of your windpipe, this vibration triggers oscillations of nearby air molecules, which then bump into their neighbours, transmitting sound waves at 330 m per second through the air. And the character of the sound produced by these vibrations is tightly controlled by the positions of the vocal cords, tongue and palate.[47]

But in birds, it all happens rather differently. Instead of using the larynx, which is located at the upper end of the windpipe, the specialist voice box in birds, the syrinx, sits at the lower end (Figure 2). Owing to this arrangement, birds vocalise from a point deeper down their respiratory tract than mammals, and it turns out that this makes them much more efficient orators.

Their longer, wider windpipes containing complete cartilaginous rings help their sounds to resonate up to their mouths, much like what happens in the whirly tube. And compared with the larynx, the syrinx requires lower air pressures to vibrate its labial folds, meaning that sound-for-sound the syringeal setup demands less energy, and pound-for-pound birds outshout all other classes of vertebrates. At only 28 cm in length and 250 g in mass, the impressive White Bellbird *Procnias albus* of the Brazilian Amazon rainforest is capable of reaching concert-topping volumes of 125 dB![48]

Moreover, by shifting the avian voice box to a point in the respiratory tract where the main windpipe divides into two, it provides the opportunity to have more than one sound source at a time. Indeed, birds have been demonstrated in experimental models to induce independent vibrations in membranes within each of these tubes, thereby enriching and diversifying the sound they can produce. Songbirds are specialists of the double melody. And the Veery, in particular, has mastered the harmonic duet.[49]

Therefore, the title of this chapter, 'The Veery's Duet', refers to a song performed not by two individuals but just one. And the syrinx is the requisite instrument. Two independently controlled pairs of vocal folds vibrate in response to two separate streams of air, perfectly tuned to execute its ethereal panpipe harmony. This prompted J.H. Langille, in *Our Birds in Their Haunts: A popular treatise on the birds of eastern North America* (1892), to describe the Veery's song as such: 'Each tone is one of many keys, all in sweet attune, a chord of many different musical threads, vibrating sweetly, and causing the atmosphere to respond as if it were itself entranced.' A fitting tribute to one of Nature's most alluring sounds.[50]

Our feathered portent

But it turns out there is much more to the Veery than meets the ear. The Veery's scientific name, *Catharus fuscescens*, derives from the Greek word 'katharos', meaning 'pure' or 'clean', and the Latin word 'fuscus', meaning 'dusky' or 'hoarse'. Given these somewhat conflicting terms, it seems that scientists have not quite known what to make of this enigmatic bird.

As a North American migrant thrush, the Veery breeds in Canada and northern USA in May–August before travelling across the Gulf of Mexico and the Caribbean Sea, thereby arriving in September–October at wintering grounds in the Amazon basin. While this inter-hemispheric, cross-continental existence is a timely symbol of the world's inherent connectedness (lest we forget), the Veery's skill as a long-haul aviator has garnered much interest among researchers.[51]

Between the 1960s and the 1990s, it was observed that in the year following particularly severe autumnal hurricanes across the Gulf of Mexico, populations of North American migratory birds such as the Veery shrank. The most obvious interpretation of the resulting storm hypothesis is that since the hurricane season coincided with the birds' autumnal migratory period, this led to greater mortality en route. However, this may not tell the whole story.

Remarkably, the timing of the Veery's breeding season seems to be a portent of upcoming weather extremes. Veeries typically fledge a single brood each spring, but it may take more than one nesting attempt to be successful. In some cases, unsuccessful Veeries must interrupt their reproductive instincts in order to prepare for their demanding southerly migration. These birds have been shown to cease their breeding season earliest in years that turn out to be the most hazardous during the hurricane season. It is as though they are forewarned of the impending danger, conserve their energy in preparation and allow more time for the gruelling journey. The inevitable consequence of these curtailed breeding efforts is a reduced population the following year.[52]

Exactly how the Veery and other long-distance migrants act as bellwethers for meteorological trends is not known. Undoubtedly, strong survival pressures in these species would have favoured those that adapted the sensory and physiological mechanisms to respond to known predictive factors of forthcoming hurricane intensity. The most relevant factors in this context include spring air temperatures in the birds' northern breeding regions, sea level pressures in the Gulf of Mexico during their northerly migration and the degree of rainfall in their southern wintering zones. By sensing and amalgam-ating such climate patterns over the full annual cycle across their extensive geographical range, the birds could feasibly build up a probabilistic picture of how their chances will fare during the next hurricane season.

Whichever way these 29 g songbirds manage to accomplish their biannual odysseys, they epitomise the intricately woven connection between biology, geography and meteorology. But with recently shifting weather patterns as a result of human-caused global warming, the strands of predictability on which so many biological systems have depended for so long are becoming weathered.

More generally, it is clear that complexity and diversity of life cycles are among Nature's best ways to build resilience. If one phase of a complex cycle becomes threatened, the remaining phases can hopefully pick up the slack while it adapts. If one species succumbs, then a biodiverse region is more likely to have another closely related species ready to expand into that niche, thereby preventing a domino effect of failing species. But this is only possible if the development of these systemic threats is relatively slow and weak.

What the relative rapidity and intensity of today's global warming will mean for species such as the Veery, or any product of Nature for that matter, is difficult to predict. There is simply no precedent for the set of circumstances we find ourselves in. As far as our observations inform us, never before has one species (*Homo sapiens*) dominated Earth to such an extent that its activities knowingly threaten its own existence.

In my hypothetical scenario in the introduction, the over-sporing New Zealand Vegetable Caterpillar fungus was only doing what came naturally to it. How could it possibly have known that prodigious short-term gains, when obtained unsustainably at the expense of its vital and limited resource, the porina moths, would result in its eventual reproductive failure?

Sitting in our ivory tower, we make the assumption that the fungus' lack of a central nervous system would have impaired its cognitive and reasoning abilities to predict what might happen. We presume that its inability to form complex global communication networks and develop accurate analytical frameworks would have precluded such foresight. We conjecture that the majority of fungi would have been so busy consuming caterpillar innards that they would have forgotten to consider what might occur if those caterpillars ran out. We speculate that had even a small minority of these fungi observed the signs of their impending ecological collapse, all the other fungi would have listened and taken heed. But then, maybe our tower is not made of strong ivory after all but is instead a brittle ceramic one.

The Veery's duet that I heard on that warm Canadian afternoon was the latest version of a song that Veeries (and their ancestors) have been performing to each other for millions of years. 'The tones ... are the sweetest I have ever heard, and can be compared to nothing else which ever falls upon the ear,' wrote J.H. Langille in 1892. Perhaps intended by the bird as a territory marker, an intruder warning or a flirtatious quip, it most certainly was not targeted at human ears. This matinée performance seemed to be a chance encounter, as so much in Nature often is. But the reality is that Nature is highly regulated with feedback loops to maintain the status quo wherever possible. Only when observed in the right context and over the appropriate timescale does this become clear. Even the cornerstone of evolution, random genetic mutation, may be more regulated than we give it credit for.

Perhaps, therefore, I can be afforded a little poetic licence and claim that, aside from the Veery song's absorbing aural qualities, this was a decisive call for help from within. It was an internal alarm, signalling unrest in the natural world; an SOS message from the defenceless to the powerful.

It was Nature's plea for salvation.

It was Mother Nature begging us to conserve Her health.

CHAPTER 2

The Façade of Knowledge

The unknown knowns are what bound us

To introduce the theme of this chapter, please join me at the Johari window. The most obvious thing about this window is that it has four panes of glass. The first pane in the top left is labelled the arena, through which both you and I observe the same view of the outside. Moving clockwise, the second pane represents my blind spot, which allows you to see the things that I cannot. The third pane represents the darkness, which stops both of us from seeing anything at all. Finally, the fourth pane in the bottom left is my façade, which hides from you the things that I can see clearly.[53]

Devised by psychologists Joseph Luft and Harrington Ingham in 1955, this model for resolving interpersonal conflict has been repurposed several times. The most famous occasion was in 2002 when the then US Secretary of Defense, Donald Rumsfeld, defended his country's actions in the prelude to the Iraq War. He baffled his audience when he spoke of the known knowns (the things we know we know), the known unknowns (the things we know we do not know) and the unknown unknowns (the things we do not know we do not know).

These are the first three panes of the Johari window, respectively: the arena, the blind spot and the darkness. But perhaps the most intriguing pane is the one that was missed out: the façade. This represents the unknown knowns. These are the things we think we know but do not.

In the continuous pursuit of scientific knowledge, these are the things science provides a glimpse of but hides the rest away. If you scrunch a piece of paper into a ball and draw a line around it, it gives the illusion of being a complete ring. Uncrumple it, and the gaps between segments reveal it is anything but. It is the unknown knowns that create the façade of knowledge.

This chapter is about our species' role as the responsible guardians of scientific knowledge. And how knowledge can just as easily misguide us as it can guide us, particularly when it comes to our relationship with the natural world.

Knowledge is both inconstant and incomplete

Science derives its name from the Latin word 'scire', meaning 'to know'. A scientist, therefore, is 'one who knows'. But what does it really mean to know something? Is it enough to regurgitate facts, or is it more of a process one must go through? How does one weigh up the validity of a piece of knowledge, and what if two bits of knowledge contradict each other? Once in existence, is knowledge set in stone?

It doesn't take a scientist to know that Christmas Day falls on 25 December. Yet it is to science we turn when we want to know the likelihood of a *white* Christmas. To know the likelihood of whether something might happen is not what many people might think of as knowledge. After all, a likelihood represents an element of the unknown, and this uncertainty is borne out as a prediction. It provides the percentage chance of a certain event occurring, in this case the chance of at least one snowflake falling during a particular 24-hour period.

In order to make this prediction, the criteria need to be strictly defined so that a record of past similar events can inform the current one. A model linking known outcomes from the past with known or suspected predictors is constructed in an attempt to 'know' something about the future.

In this example, 100 years of data might be used to link the history of snowfall on Christmas Day with observed weather patterns during the days, weeks or months beforehand. The confidence scientists attribute to any particular model is bolstered when larger volumes of good quality data are used to build it and when there is a shorter time interval between the prediction and the predicted outcome. These serve to reduce the error of the model, thereby increasing its accuracy.

But while models help scientists understand the unknown bits of science, there are some things that today's scientists know for certain.

This includes the fact that humans typically have 46 chromosomes in each of their cells. There are notable exceptions, of course. For example, the sex cells have half that number, and red blood cells, which account for 84% of our 30 trillion cells, exchange their chromosome-bearing nucleus for greater oxygen-carrying capacity.[54]

Nevertheless, humanity's knowledge of chromosomes has been nothing short of revolutionary. This is because chromosomes contain deoxyribonucleic acid (DNA), which is the recipe book for making an organism, whether it be a Common Alder, a Veery or a human. Put simply, DNA holds the instructions for making the proteins that life needs to exist.

Therefore, considering DNA's importance to all known life, it is worth exploring how our knowledge of it came about.

Many consider James Watson and Francis Crick to be the discoverers of DNA, the phosphate-rich nucleic acid that makes up each chromosome. But in reality, their pivotal work in ascertaining the double-helix structure of DNA, published in 1953, relied heavily on previous work by others.[55]

The first descriptions of what he called nuclein were made by Friedrich Miescher in 1869, documented in letters to his anatomy professor, Wilhelm His Sr. Miescher explains how he took pus-ridden bandages from the neighbouring clinic in Tübingen, Germany, and devised a method for isolating the pus cells, also known as white blood cells. As a student of chemistry, Miescher was most interested in the components that made up the nucleus of the cell, which is notably large in white blood cells. The nucleus was of topical interest in the scientific community after Gregor Mendel had recently proposed his laws of inheritable traits in peas and Ernst Haeckel had then suggested the nucleus might contain the substance for such inheritance.[56]

Despite his curiosity in the underlying chemistry, the extraction of white blood cells from soiled bandages in the pre-antibiotic era must have been a grim task. Indeed, Miescher noted in his letters to His that the ones revealing 'further decomposition by sight and smell' were discarded.[57]

While his level of devotion was unquestionable, his discovery of what would ultimately come to be known as DNA was met with scepticism by his supervisor. And although his work was published in 1871, its reception among the wider scientific community was muted. Nonetheless, over the following 82 years, chromosomes, genes (specific sequences of DNA that instruct for particular proteins), bases (fundamental units of DNA code), mutations (errors in the DNA code) and gene–protein relationships were all described, well before Watson and Crick publicised their double-helix DNA structure in 1953. At what stage would it have been justified to say scientists 'knew' about DNA? Even now, over two decades since the first human genome was fully sequenced, do we know everything there is to know about DNA?[58]

Therein lies the issue with knowledge. It is inherently mercurial, a product of time, people and circumstance. But it is also cumulative, relying on curiosity, persistence and luck to reach a point where a consensus is formed. The unknown becomes the known, even if it is almost impossible, in most cases, to pinpoint when this transformation occurs.

Nobody in the scientific community today doubts the existence of DNA, as the amassed evidence in the literature speaks plainly. Despite this, how many of today's scientists have recreated the experiments to prove it first hand? I know I haven't. There is trust in the experts that what gets reported in peer-reviewed journals is bona fide. This allows the stream of knowledge to continue its flow. Today, constructive debates rage on among scientists as to the relevance of certain mutations in specific diseases or the mechanisms by which certain genetic defects cause cellular dysfunction.

These are the currently unknown parts of the known that set the research agenda while keeping scientists humble.

Learning by example

The neurological disease I have studied most closely is motor neuron disease, also known as amyotrophic lateral sclerosis. This is a devastating illness that takes away the abilities humans tend to take for granted: the ability to walk, write, speak, swallow and breathe. The finely tuned network of nerves that grows and flourishes in health withers in disease. Impulses that once had the know-how to travel at spine-tingling speeds of 100 m per second are forced to slam on the brakes. And muscles fall victim to a relentless and terrible siege. Tragically, a half of patients die within two years of diagnosis.[59]

By looking at the contents of affected nerves during post-mortem evaluation, it has been observed that 97% of patients with this disease have accumulations of a specific protein called TDP-43 (standing for the awkward but descriptive term 'transactive response DNA-binding protein of 43 kilodaltons'). The most tempting conclusion to be drawn from this association is that the build-up of this protein somehow blocks the normal cellular functioning within the neuron (the technical term for a nerve cell), thereby leading to its demise.[60]

However, association does not equal causation. It is perfectly feasible that this observation is a bystander effect of the true cause. This would be like leaving your car lights on overnight and then blaming the lightbulbs when your car doesn't start the next morning.

Two other observations make this scenario more perplexing. First, the aggregated TDP-43 protein is present in patients who inherit a causative genetic mutation as well as in patients who develop the disease without a known mutation. Second, the 3% of patients without a build-up of TDP-43 are clinically indistinguishable from everyone else, despite accumulating a different protein inside their neurons.[61]

Going back to the car analogy, the first observation is akin to cars from a single factory failing in just the same way as cars made in many different factories across the world. The second observation is like realising it doesn't matter whether it is the outside or the inside lights that are left on; the car does not start just the same.

The trouble is that scientists have not yet been able to identify the neuronal equivalent of the car's exhausted battery.

And this incomplete understanding of the disease prohibits effective therapy. Like echoless bats in search for food, scientists have been flapping frantically in the dark for a cure. In the mid-1990s, the drug riluzole offered a tantalising glimpse of a therapeutic bounty awaiting discovery. But after three decades of clinical trials, an Ice Bucket Challenge and hundreds of failed drugs, patients remain perched on a crest of false hope. While a recent wave of personalised treatments targeting specific genetic mutations has rightly injected renewed optimism, only time will unravel the true impact of these innovative therapies.[62]

The partly genetic basis of motor neuron disease was previously unknown owing to limitations in even the basic understanding of DNA. But thanks to the pioneering work of Miescher, Watson, Crick and countless others, current knowledge is creating opportunities for therapeutic progress in this dreadful and incurable neurological disease.

In one sense, the progress of knowledge appears that it knows no bounds, but at the same time it would be our collective arrogance to underestimate the power of the unknown to keep scientists in check every step of the way.

Knowledge is a tree

Knowledge grows from the seed of a hypothesis. This is a speculative idea that demands, in the mind of the hypothesiser, experimentation within the observable world to confirm or refute it. Miescher's idea in 1869 was that the contents of the cell's nucleus may explain one of the pertinent biochemical questions of the time. This concerned the mechanism for inheritable traits, an idea of its own put forward by Mendel four years earlier. Miescher had the necessary inquisitiveness, motivation and doggedness to find out which chemicals were in the nucleus of white blood cells. His discovery of an unfamiliar substance that behaved differently to any biological substances known at that time vindicated his efforts, although he could hardly have predicted the full ramifications of his breakthrough.[63]

Nestled underground, a tree seed first sprouts its rudimentary root network before its stem breaches the soil surface as a seedling. Its continued growth in the presence of sufficient sunlight, water and nutrients converts this seedling into a sapling, eventually maturing into a fruiting tree after several years, sometimes decades. Was Miescher's discovery of nuclein the sprout of a new seed or a new branch on a seedling, sapling or mature tree? Or maybe it was all these things simultaneously?

In just the same way that a seedling is susceptible to irreversible injury from a trampling hoof or a hungry fawn, early rejection and denial can squash new knowledge before it ever gets going. Enthusiasm and corroboration are its sunlight and water, while healthy scepticism and constructive critique are the perfect mix of nutrients to support growth.

Reproducibility and external validation of the results are the subterranean fungal network that forms the facilitatory wood wide web. A contradictory set of results may be the lightning strike that fells an established tree, or it may simply lop a few branches off, creating an opportunity to grow again in a revised fashion. Saplings may grow in the shade of surrounding stalwarts, thereby starved of the sustenance needed to flourish. Some will bide their time long enough, but others will wilt, wither and withdraw.

Furthermore, a single tree can be viewed from varying angles or in different seasons and attract a range of descriptions, underlining the whims of scientists when presented with the same set of results. In many cases, the growth of new trees may be purposefully hidden from wider view in the same way that negative results may be considered too bland for publication.

I want to draw one last parallel between the healthy growth of trees and the accumulation of knowledge. But first, we need to understand that a tree, for the most part, is the ordered assembly of just three things: carbon, hydrogen and oxygen. In fact, these three elements make up over 90% of a tree's dry biomass. At this elementary level, there is no material difference between the emergence of the first sprout from an acorn and the terminal branch of a 600-year-old oak bough.[64]

This is how Miescher's nuclein discovery, comprising its three ingredients of method, observation and interpretation, could represent both the seed *and* a new branch on the tree of knowledge. It simply depends on the timescale and context you view it from.

Biological warfare

We can take the analogy between trees and knowledge even further by looking at their shared vulnerability to attack.

In the summer of 2020, citizen science in Canada captured the first glimpses of an insidious invasion. Snaking its merry way from the edge of an elm leaf, the caterpillar-like larva of the Elm Zigzag Sawfly *Aproceros leucopoda* consumes everything in its path, leaving nothing but the leaf's greenless skeleton after just two weeks. Evolution of this species found no need for males, and so females reproduce asexually, turning the monotony of mating into a maternal monopoly. This significantly shortens the sawfly's reproductive cycle, and metamorphosis from larva to adult is rapid, occurring in just three to four days within its cocoon. All of this means that four to six generations can appear during the warmer months. And this eruption of sawflies proceeds to ravage elm trees from crown to trunk.[65]

Unfortunately, this is far from the only attack confronted by native elm trees in Canada. Substituting the fabled Trojan horse for bark beetles and its hidden Greek warriors for a specific fungus, Dutch elm disease has decimated elm numbers by as much as 80% since the 1940s. It does this by clogging up the tree's water canals, inducing an internal drought. Meanwhile, in a different disease known as elm yellows, certain bacteria parasitise the tree's sugar-transporting channels after being spread from tree to tree by sap-loving insects known as leafhoppers. Ultimately, these diseases can lead to the death of susceptible elm trees within just a few years.[66]

Whether insect, fungus or bacterium, the native elms in North America (and other areas of the world such as Europe) are vulnerable to these alien blights because geography had kept them apart until very recently. Only through human global colonisation have these organisms met, but it has hardly been a match made in heaven.

In the introduction, I wrote about how the Vegetable Caterpillar fungus interacts with the porina moth as part of its natural lifecycle. And I speculated what might happen if this relationship were to become an unequal, unsustainable one. Here, though, we have real-life examples of the domination of one species (the parasite) over another (the elm tree). But where are the natural checks that should be keeping these biological relationships in balance?

The answer epitomises the process of evolution. In areas of Asia where these parasites are native, indigenous elms tolerate their presence without suffering major damage. After millions of years of co-exposure, mutual mechanisms exist to ensure the parasite does not overcome the host.

For example, in naturally resistant elm trees, water vessels are smaller and more widely spaced, limiting spread of the fungus that causes Dutch elm disease. Moreover, inoculation induces a chemical riposte from the elm that serves to compartmentalise the fungal invader.[67]

And in reaching Europe and North America, the Elm Zigzag Sawfly operates beyond the native range of its natural predators, effectively loosening the shackles that would usually restrict its population growth. Regrettably, it's the European and North American elm trees that fall victim to the sawfly's overpopulation.

These examples provide a unifying message. Host and parasite are subject to a complex ecological interchange that serves to stabilise both populations. But recent unexpected encounters have given rise to unstable exchanges. These abrupt meetings are the direct result of a deeply connected world, as human activities dismantle long-established barriers between distant biological rivals. Each parasite is only doing what it has perfected on home soil for thousands of millennia. But suddenly, among ill-equipped hosts on away turf, its usual resistance capitulates.

A communications failure (of sorts)

This draws parallels with the way scientific knowledge is accessed, interpreted and disseminated in our modern, more connected world. Scientists spend decades learning how to formulate the most relevant hypotheses, construct the most rigorous methods and perform the most appropriate analyses. At the end of all that, interpretation is rarely binary. Studies often generate more questions than answers, and the Ferris wheel of scientific knowledge resets and cycles around once again. At times, science is more of an art than a... science.

There is simply a cavalcade of obstacles over which even the wariest may stumble. Nuance, loopholes, caveats and pitfalls – these are all out there to hoodwink both the naïve and the learned. Except, those with greater experience naturally have their ears pricked and eyes peeled for these potential dangers.

For example, biases can occur when selecting participants for a study without randomisation, or when grading outcome measures without being blinded to the study groups. If the selection criteria are too lax, the high variability within the study group will make it trickier to detect a positive outcome. But if the selection criteria are too tight, any positive results lack generalisability within the wider population.

Furthermore, a false-positive result can occur if too many outcome measures are unwisely applied, and one of them detects a positive outcome just by chance. Meanwhile, a false-negative result may happen if insufficient participants are included in the study, and the outcome measure misses the true effect being investigated.

So how do scientists navigate these obstacles? They defer to the guiding light of statistics. The Scottish poet and novelist, Andrew Lang, is credited with saying that 'most people use statistics like a drunk man uses a lamppost; more for support than illumination'. I might add that often it simply gets urinated on.

Despite this, statistics governs the methodology, analysis and interpretation of science. Statistics guides how many participants to include in a study. It provides a way of dealing with the inherent variability of a sample. It tells us how certain we can be that any given result reflects reality. Perhaps most crucially of all, statistics sets the benchmarks for acceptable rates of failure.

Typically, a false-positive is permissible up to 5% of the time, while a false-negative is tolerated up to 20% of the time. Given the sheer number of scientific experiments conducted across multiple disciplines, this amounts to a lot of legitimately wrong results. But it is kept to a tolerable level, much like the ability of Asian elms to keep native parasites at bay. Deemed unavoidable, this misinformation contributes to the *reputable* side of the façade of knowledge.[68]

Now consider what happens when these sources of error are mismanaged. After all, the diligence of the scientific method exists solely so that scientists can judge the robustness of their results and make appropriate recommendations to peers, policymakers, governments and the wider public. When the scientific method is neglected, either through ignorance, ineptitude or infamy, the spread of misinformation balloons beyond acceptable levels.

We know Canadian elms are unable to defend themselves against the unfamiliar Asian fungal pathogen that causes Dutch elm disease. The ability of the fungus to spread through the elm's vital water channels is analogous to the ease of modern global communication. The knowledge acquired through neglect of the scientific method clogs up and contaminates the long-established flow of knowledge that *does* adhere to the scientific method. Further insults from destructive sawflies and bacteria eventually topple the elm. Harmfully, this type of misinformation creates the *disreputable* side of the façade of knowledge.

At its core, science is about observing the world and forming plausible stories that best explain those observations. It is in the domain of anyone with an imagination and a desire to know more. It can be a hobby or a profession; it can be carried out by the young or the old; it can tackle simple questions or more complicated ones; the answer could come in an afternoon or take several decades. A scientist is any responsible practitioner of a guiding set of scientific codes and ethics; a scientist is anyone willing to devote their relatively short time on Earth to the relentlessly growing body of scientific knowledge. But there are few shortcuts, and ulterior motives only serve to misguide. Like the growth of an elm tree, good science takes time. And as valid evidence amasses, what started off as the plausible might become the probable, before firmly taking root as the indisputable.

Nurturing our models of the world

The air cools. The hairs on your forearm stand to attention, as a chill sweeps through your core. A dark cloud curtain with distant rain tassels creeps towards you. You curse the inaccuracy of the morning's forecast as you stand on a country path ill-prepared for inclement weather. Instinctively, you scour your surroundings for cover, as the first waterdrop bounces off your brow. The warmth of the sun from moments before fades into an intangible memory. You can smell the stirring squall. You hear the faint rumble of thunder.

Two trees stand grandly nearby. One is broad with shiny, lobulated leaves; the other is tall with flat, soft needles. You peer under each canopy to confirm that each tree gives the illusion of complete cover from the elements. You attempt to compute the relative chances of staying dry under each one. You suspect the proportion of rain either caught by the leaves or directed towards the trunk will be affected by tree height, canopy breadth and branch structure, as well as the density, size and angles of the leaves. You wonder whether standing near the trunk or at the edge of the crown is optimal. You stand there contemplating the flow of raindrops splashing off each leaf, streaming down each branch and sliding through the gaps. If it just turns out to be a short downpour, perhaps the taller tree with more branch layers will be more protective. But if it turns out to last longer, then perhaps the taller tree will produce larger and faster raindrops once they eventually reach you.

Just as you conclude that your internal model of how rain travels through various tree canopies is hugely deficient, you see a flash brighten the sky, shortly followed by a protracted grumble. You take this new information on board and realise standing under a tree would be unwise after all. And you decide to head for lower ground, getting soaked in the process.

A model is a way of simplifying the complexity of life. Inevitably, a model is an incomplete view of reality, steeped with generalisations and assumptions that can lead to incorrect predictions and inefficient decision-making. Indeed, a model is only as good as the empirical observations used to construct it. As new information is assimilated, priorities can change and what once seemed important may be discarded. In the scenario here, the model forecasting the weather for that locale had not predicted the spontaneous storm. The more widespread weather patterns used to formulate the morning's predictive model allowed focal anomalies such as this to be missed.

And while models of rainfall partitioning into throughfall (rain passing through the gaps in foliage), stemflow (rain passing down branches and trunks) and interception (rainfall that evaporates directly from leaves) are important in evaluating water cycles, soil erosion and nutrient flux in the science of silviculture, they are not so helpful in our everyday lives. The improbable nature of this sequence of events renders the usefulness of any hastily constructed internal models imprecise.[69]

It turns out that our scientific models thrive off the unknown knowns. The unknown knowns are what prompt scientists to nurture their models over time, as they update what they thought they knew but now know differently.

Take animal models of human disease as an example. Whatever your stance may be on the ethics of such work, there is no doubt that significant breakthroughs have been made in many areas of medicine, from antibiotics to chemotherapy, through the use of animals in research. Cast in bronze and squat in scholarly repose, a mouse knits the double-helix structure of DNA. On display outside the Institute of Cytology and Genetics in Siberia, Alexei Agrikolyansky's *Monument to the Laboratory Mouse* fittingly commemorates the tireless use of animals in the pursuit of science.

Thankfully, scientists involved in animal research today adhere to strict ethical approvals and licensing processes to ensure animals are treated as humanely as possible. The purpose of this work is to provide a model of human disease that can be subject to manipulation and experimentation in ways that would be wholly unethical in humans. While experiments on flies might pass most people's ethical gauge of acceptability, the involvement of non-human primates naturally raises greater anthropomorphic concerns.[70]

There is a well-studied murine model of multiple sclerosis, induced by the injection of proteins that are detected as foreign by the mouse's immune system. These proteins are associated with the fatty substance that encases neurons, known as myelin. In multiple sclerosis, myelin sheaths are attacked by the patient's own immune system, and the mouse model has been artificially designed to simulate this uniquely human process. Mice do not naturally get multiple sclerosis, and therein lies the problem. Even if scientists find an effective treatment for the artificial animal model, there is absolutely no certainty that the treatment will be equally effective in the real human disease. The façade of knowledge is an imposing obstacle.[71]

Fortunately, this mouse model of multiple sclerosis recapitulates the human disease quite well, or at least partly. It was the driving force behind the development of many drugs that have been successful in human clinical trials. These drugs reduce the rate at which patients experience relapses, but there is a crucial caveat. These drugs do not appear to stop the progressive phase of the disease that manifests many years after diagnosis. But this should come as little surprise given that the mouse model was only engineered to imitate the inflammatory process during relapses, not the degenerative process that follows. It highlights the reality that a one-size-fits-all model is rare.[72]

Often, there are obvious discrepancies between the model and the modelled from the outset. The most common form of dementia worldwide, Alzheimer's disease, has been modelled extensively in mice. In humans, the disease is associated with plaques of amyloid outside neurons and tangles of tau within cells. But while the classical mouse model shares the amyloid hallmark of the human disease, it omits the other. And this creates an obvious problem for scientists when it comes to translating their findings from mouse to human.[73]

Similarly, any attempt to model human motor neuron disease in mice is influenced by inherent differences in the neuronal pathways between the two species. From brain to muscle, humans have evolved a dextrous two-cell relay, whereas mice rely on a more circuitous path.[74]

What these examples indicate is that any extrapolation from model to human is a perilous task.

Models come in all shapes and sizes

Models can be biological, chemical, computational, conceptual, linguistic, mathematical, physical or stochastic. They can tackle an overarching theme or focus on a niche. They can require symposia and consortia to fully flesh out, or they can be written on a napkin in a late-night bar. They can stand the test of time or be a flash in the pan.

But despite this variety, all models have something in common; they are an imprint of the true object in question. Some are more consistent representations than others, but all are creatively constructed for their explanatory and predictive power.

Perhaps the closest a model can get to the real thing is a natural history study. This is a model that plays out in a natural setting, mitigating the detachment all models suffer from to some extent. Let me share two examples of what I mean.

The House Sparrow *Passer domesticus* is one of the most numerous and expansive perching birds in the world. The key to this species' success has been its reliance on humans. As urbanisation of our own species has exploded, so too have populations of House Sparrows. However, one wouldn't need to time-travel too far back to discover a conspicuous absence of these anthropophilic birds in many parts of the world.[75]

In 1851, eight pairs of House Sparrows were introduced in New York to help Europeanise the Americas. What happened next serves as a neat model of biological adaptation.

When researchers today look at variations in bird size and colour across North America, a clear correlation presents itself. Larger, darker House Sparrows are preferentially found in the colder northern territories, while smaller, paler birds are found in the warmer southern climes. Since larger birds have lower surface area to body mass ratios, they conserve heat more efficiently. Meanwhile, darker birds absorb more of the sun's radiation. Therefore, a larger, darker House Sparrow is not only more likely to survive in colder average temperatures but is also more likely to overheat when average temperatures are warmer. The opposite is true for smaller, paler birds, leading to the observable spectrum of these attributes across North America.[76]

This all happened in 100 years or so, testament to the strong selection pressures placed upon these birds. This is an accelerated example of selective adaptation. It is a model of natural evolution on fast-forward.

For my second example of a natural history study, I turn to work I carried out during my PhD. One of my main objectives was to devise and test a new way of counting spontaneous muscle twitches in patients with motor neuron disease. Termed 'fasciculations', these twitches are known to reflect increased excitability of affected neurons, as though it's their last call for help before succumbing to the disease. So universal are fasciculations in the relatively early stages of disease that neurologists are very wary of making a diagnosis of motor neuron disease in their absence.[77]

While fasciculations can often be seen with the naked eye, their electrical component has traditionally been detected by sticking a fine needle into affected muscles. But more recently, this can also be achieved by sensors that stick to the skin surface. For now, this non-invasive, better tolerated approach remains a research tool. And it was this method that I used during my PhD.[78]

I wanted to build up a model of what was happening to fasciculations as patients progressed naturally through the disease. Without a meaningfully effective treatment, this was an approach that would hopefully lead to a better understanding of the underlying excitability of nerves over time. And this knowledge might lead to new therapeutic discoveries.

With the help of bioengineer collaborators, I developed an automated system for counting and characterising fasciculations from electrical sensors placed on the patient's skin. This provided information such as fasciculation number, size and firing pattern. This automated algorithm enabled me to apply the same analysis to large quantities of data, limiting the bias that can occur during manual forms of analysis. I invited 25 patients to attend King's College Hospital in South London every two months for a year, recording four muscles each time.[79]

Although this was a natural history study, designed to be the closest possible model of the real disease, there remained recognisable problems. Generally, patients who agreed to join the study were progressing more slowly than average, making them a skewed representation of the wider patient population. And this only worsened as the study progressed, given that the most affected patients were more likely to drop out.[80]

Additionally, I was only taking infrequent snapshots of fasciculation behaviour (once every two months), and I could only guess at what was happening between visits. Plus, I only had the time and resources to assess four limb muscles, only a small proportion of the many hundreds of muscles in the human body.

Added to these issues, there were problems with artefacts appearing in the data, as even the presence of laboratory lighting could interfere with the minute electrical signals I was trying to detect.

The fact that patients had to spend considerable time travelling to a busy London hospital was important, too. The increased exertion, the sleepless night beforehand, the skipping of breakfast, the coffee on the train and the stress of a congested carriage could have all influenced the fasciculation measurements.

Although you try as a scientist to control for these variables as much as possible, the study of human disease is fraught with such challenges. You are interacting with patients at the ebb of their health, all contributing altruistically in the knowledge that studies such as these will not generate breakthroughs quickly enough for their personal benefit. Admirably, they contribute purely in the hope that someone else in the future will not be forced to endure a similar fate. These patients plant trees whose shade only others will sit in.

Scientists the world over strive to overcome the inevitable imprecision of disease models. In the case of my PhD, I did manage to develop a unifying model linking the excitability, loss and reorganisation of nerves. But this is something that other scientists and I continue to tweak. With each study, we aim to construct a more refined model of disease that can be applied to test new therapies. For the patients involved in these studies, a therapeutic breakthrough cannot come soon enough.[81]

Taking the strain

While the focus of this chapter has undoubtedly been on knowledge, I am the first to admit that knowledge on its own only gets us so far. Instead, it is how we apply that knowledge to fix our worldly problems that really matters. Therefore, I now turn to an approach that helps us think about our problems in a general sense. And, more importantly, how we might go about fixing them.

At the heart of any problem is a system, whether it be the human nervous system or a Veery's ecosystem. And a stressor is anything that puts strain on that system. The system may absorb the strain, showing resilience (problem solved), or it may fail, showing vulnerability. But given the tools and time to do so, a pliable and sustainable system adapts itself to better accommodate future impacts from the same stressor.

Let me elaborate with a visual model (Figure 3).

Imagine a volcano-shaped fluid vessel. On its sides, it has multiple spouts of varying shapes, sizes and orientations. A storage tank is

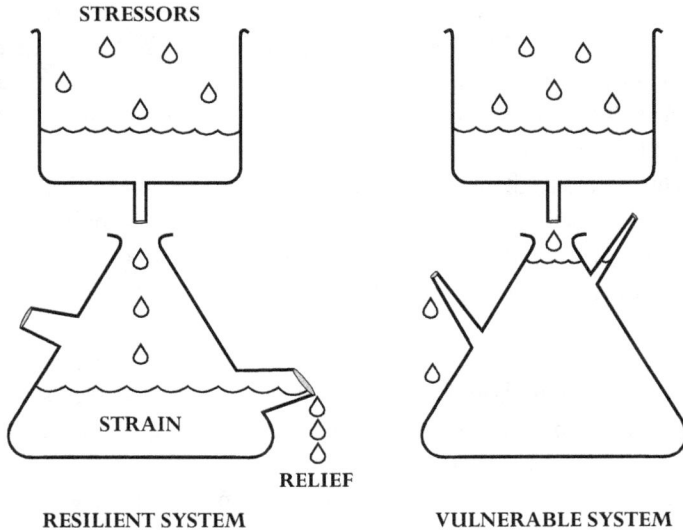

Figure 3. The stressful vessel model: on the left, a resilient system continuously relieves the strain and maintains it at sustainable levels. On the right, inefficient coping strategies lead to unsustainable levels of strain on a vulnerable system.

suspended above it. Fluid pours into the tank, drips into the vessel and drains through the spouts. If the rate of filling in the vessel exceeds the rate of emptying, the fluid level rises. Initially, owing to the wider diameter at the base of the vessel, the rate of this rise will be slow. However, as filling continues to outstrip emptying, the faster the fluid approaches the top. Overflow represents a compromised system, a point of irreversible failure.

Now, the overall systemic strain caused by all stressors is represented by the height of the fluid in the main vessel. And the spouts represent the system's coping strategies. Spouts with shallower angles, wider channels and shorter lengths drain the fluid most efficiently, thereby providing faster relief. And the problem is promptly fixed.

The main concept here is that each drip from the storage tank into the main vessel stresses the system in some way. But this is not a new idea. This is similar to the stress bucket model introduced by Alison Brabban and Douglas Turkington in 2002 to conceptualise the stressors that trigger mental ill health. However, the stressful vessel model that I present here emphasises two additional points.[82]

The first is that there is a variable lag between the occurrence of the stressful event (fluid added to the storage tank) and the strain it puts on the system (fluid dripping into the main vessel). In some instances, the lag will be very short, almost instantaneous, but very often it is more drawn out and insidious.

This serves to demonstrate an oft-forgotten principle. Even when a stress-inducing event is ostensibly over, it can continue to cause problems for some time. The fallout from this is that it can be difficult, almost impossible, to form a clear temporal link between a stressor and the resulting strain.

The second concept emerging from this model is one of adaptability. This can be thought of as a flexible response to a fixed stressor. It represents coping, and natural systems are excellent at it. Some coping strategies are generalists, enabling the system to deal with a broad range of stressors; others are focused on a specific task.

These adaptations are represented in the stressful vessel model by the spouts. Those nearer the bottom represent core coping methods that enable the impacts of stressors to be kept to a low level. Spouts higher up the main vessel are contingency strategies that become vital if other spouts become blocked. Where a spout has its tip above the top of the main vessel, this represents a misguided strategy that only serves to use up limited resources without ever preventing systemic failure.

The overarching point is that problems are best fixed by a system that is able to adapt over variable timescales.

Putting this model into practice

To exemplify the utility of this stressful vessel model, let's reconsider the problems facing the elm tree. Earlier in this chapter, I discussed how the elm tree has variable levels of susceptibility to the same stressors depending on its ancestry. Asian elms are resilient to the parasites they have co-evolved with over millions of years, while Canadian elms are helplessly susceptible to these previously unseen arrivals.[83]

In the case of the Asian elm, it has evolved general traits that enable it to fend off an attack on its sap-flowing canals, including that from the fungal pathogen causing Dutch elm disease. For example, these sap-transporting channels are narrow, widely spaced and packed full of cellulose, providing structural barriers to the intruder. This generalised form of resistance is modelled as a spout near the base of the main vessel. It begins to drain the fluid at low levels, analogous to the elm tolerating a larger burden of fungal invasion.

But the Asian elm also has an array of inducible defence mechanisms against the fungus, including the release of fungicides and substances that occlude local water channels. These effective measures contain the spread of the fungus and are modelled by short, wide-calibre spouts that serve to drain any surplus strain on the system.[84]

In contrast, the vulnerable Canadian elms do not have these coping strategies, as they never had the need for them. The selection pressure from repeated fungal invasions was never present to bring about their evolution. In the place of efficiently draining spouts near the base, all that these trees can muster are long, narrow-calibre spouts that extend almost vertically. One inefficient coping strategy is the dieback–regrowth cycle that plays out in the most affected parts of the tree, capable of delaying tree death, but incapable of averting it. These are resource-intensive methods when the strain on the system is already high. No wonder they prove to be ineffective.

What is apparent from resistant elm trees is that the best systems have a combination of both fixed and adaptable coping strategies, providing both baseline and honed responses.

While a fixed spout in the stressful vessel model is easy to imagine, the portrayal of a dynamic set of inducible spouts, able to sprout, recede and transform at will, is admittedly more tenuous. Essentially, what this adaptability depicts is a form of compensation. And for another suitably understood biological example, I return to motor neurons.

If a human motor neuron dies, it retracts from the muscle. Muscle fibres that had once received electrical signals from this now defunct motor neuron become isolated, orphaned from their trusted guide and nurturer.

Remarkably, surviving motor neurons receive chemical signals to sprout new branches in order to 'adopt' the orphaned muscle fibres. This maintains a full complement of functional muscle fibres, albeit with a reduced motor neuron pool. Consequently, muscle strength is prioritised at the slight expense of fine control.[85]

This compensatory mechanism works well when the cause of the motor neuron loss is minimal and self-limiting, such as nerve impingement at the spinal level. However, when motor neurons continue to die, as is the case in motor neuron disease, this compensatory mechanism becomes overwhelmed.

Clearly, even the best coping strategies have their limits.

The cultivation of science

To sum up this chapter, I make reference to something Sue Stuart-Smith says in her book *The Well Gardened Mind*: 'To look after a garden involves a kind of *getting to know* that is somehow always in process. It entails refining and developing an understanding of what works and what does not. You have to build a relationship with the place in its entirety – its climate, its soil, and the plants growing within it.'[86]

For me, the cultivation of science is just the same. The generation of knowledge feeds into ever more diverse, but always imperfect, models of the world around us. This, in turn, fertilises new ideas, which are supported to grow in a range of environments by anybody who has patiently acquired the appropriate tools for the job. As the seasons unfurl their idiosyncrasies, the scientist-turned-gardener learns, adapts and strategises. Storms may bluster, wounds may fester and influxes may pester. And yet, in assailing these challenges and appraising their impact, the world becomes just a little more predictable.

The greatest fallacy is not in not knowing, but in thinking nothing is unknown. And how this applies to our perception of health and disease is the topic of Chapter 3.

CHAPTER 3

Chameleons in the Clinic

Healthcare aims to salvage a state of ease now lost

On the evening of 15 April 2020, I crossed the threshold of St George's Hospital in South London. I was turning up for a night shift, something I had done countless times across six London hospitals since qualifying as a doctor in 2010. But this was no ordinary night shift.

I made my way through the long hospital corridors, which appeared even quieter than usual, even eery. Universal precautionary measures were in place: surgical masks, plastic aprons, mandatory scrubs and temperature checks. Staff were guided by a traffic light system of accessible areas: green for clean, red for infected. The friends and family of patients were excluded, no exceptions. The majority of hospital services had been shut down, freeing unprecedented levels of staff and resources to help alleviate the new healthcare burden. This was the first shift of my redeployment to the COVID-19 critical care unit.

Confirmed cases of this novel coronavirus had reached 1,918,138 worldwide that day, which had resulted so far in 123,126 deaths. The virus had reached 213 countries and territories.[87]

As I had spent four months working in a critical care unit seven years previously, this qualified me, alongside many of my colleagues, for an intensive five-day revision course in caring for the critically ill. How to optimise breathing machine settings; how to puncture a life-threatening rise in chest pressure; how to revive a slow heart with pacing pads; how to correct potentially lethal blood salt imbalances. These things seemed familiar in theory, but my relative inexperience in practice was demonstrable. These were simply not things that a neurology doctor did.[88]

In the worst case scenario, this hospital alone predicted up to 600 beds would be required for the critically ill, a 20-fold increase from baseline. Wards usually set up for the most well patients were being converted into suitable locations for the most unwell. This meant ensuring there were adequate oxygen supplies, plentiful breathing and blood-filtering machines, as well as sufficiently qualified healthcare professionals. In a matter of weeks, the hospital metamorphosed.

Healthcare systems all over the world were adapting as best they could to combat this novel global attack on human health. They had little choice. As each day passed, knowledge of the disease increased, but it would take many months before an effective vaccination was available. In the meantime, the belt and braces approach of a global lockdown was necessary to limit spread of the disease. In a very short time, the healthcare arena had altered beyond any semblance of its former self.[89]

At the start of my night shift on 15 April, the day team handed over the current patients on the critical care unit, outlining the key medical events over the previous 12 hours. New admissions were discussed in detail. The sickest of the sick were highlighted for special attention. Any essential jobs outstanding from the busy day shift were delegated. Individualised patient plans for the next 12 hours were formulated. I had once been taught that the basic aim of any night shift is to keep each patient on your watch alive. The night was a time for treading water, not for any major interventions that could wait until morning's light.

The only difficulty was that even the most experienced critical care doctors did not know how to stabilise the most affected COVID-19 patients. Many patients deteriorated so quickly into multi-organ failure that modern medicine and a limited set of resources could not avert a fatal outcome. The virus causing COVID-19 made lungs too stiff to ventilate with a breathing machine. The virus impaired kidney function so readily that there weren't enough kidney replacement machines to go around. The virus caused blood to stick together more easily, leading to strokes in the brain and clots in the lungs. At that time, it seemed there was sadly only one destination for most patients on the unit: the mortuary. It was by far the scariest period of my short medical career.[90]

But there was a sense among the healthcare professionals I worked with that we would get through it. For every unfortunate patient whom we were unable to save, we hoped our efforts would allow a more fortunate patient to pull through. For every hazmat suit we donned, for each hour we wore claustrophobic face guards and goggles on our ward rounds, for each time we carefully flipped unconscious patients onto their chests to aid lung function, we edged more closely to reclaiming control over this previously unencountered infectious disease.

This chapter is all about this delicate balance between health and disease, and how the perceptions of just one species influence all life on this planet.

The nature of disease is changing

The word 'disease' stems from the early fourteenth century, when it was used to describe a general lack of ease or comfort ('dis-' + '-ease', from the old French word 'desaise'). Its modern usage in specific reference to any disorder disturbing normal biological function came later that century. Viewed in this way, a disease is simply anything that removes the comfort enjoyed in health.[91]

During the twentieth century, the most advanced healthcare systems finally overthrew the ancient dominance of infectious diseases. Diphtheria cases in the UK reached almost 70,000 in 1934. But owing to the widespread introduction of the childhood diphtheria vaccine in the 1940s, this figure faded to just 173 in 1954. Tuberculosis, widely considered to be the oldest documented infectious disease, has been plaguing human civilisations for millennia. Testament to its longevity, it has acquired inventive synonyms along the way, including scrofula, phthisis and consumption. But an effective vaccine and multiple pharmacological treatments against tuberculosis have now largely confined this disease to individuals with an already weakened immune system. Furthermore, smallpox was the first infectious disease to be eradicated from humans in 1980, while polio is close to becoming the second. Global cases of polio have plummeted from 350,000 in 1988 to just 6 in 2021. These achievements are true marvels of modern medicine.[92]

This medical revolution has contributed to the surge in life expectancy currently enjoyed in both developed and developing countries. Since 2017, humans over 64 years old have for the first time outnumbered those under 5 years. We can now expect many more years of health than citizens living just 100 years ago. But on the flip side, this has shifted the nature of disease facing our species.[93]

Based on the word's original meaning, disease converts one who is at ease into one at dis-ease. Where those at dis-ease meet one who might be able to ease their plight, a transaction takes place. The healthcare setting triangulates the disease, the dis-eased and the easer. The service sought and delivered is intimate, fretful and consequential. There must be trust, transparency and truthfulness. The prerequisite for a positive outcome is a shared appreciation concerning the nature of the dis-ease: its causes, its effects and its timespan. The prevailing knowledge at the time and place of this transaction dictates how the dis-ease may best be alleviated, if possible. Communicating this message is pivotal. Purposeful inaction can be just as powerful as evidence-based action, for time is just a mechanism for natural processes to play their healing hand.

However, I have already discussed in Chapter 2 how unknown knowns and imperfect models create the façade of knowledge. So each step forward requires caution. Undoubtedly, there are sometimes quick fixes in medicine, but increasingly the problems faced by patients and their doctors get harder and harder. Logic suggests that anything easier to solve than the current insoluble problem you face has probably been solved already. Modern medicine has become a victim of its own success.

As I have already alluded to, we have now entered the age of ageing. Heart disease, stroke, dementia and many cancers were not problematic diseases until recently. Humans simply did not live long enough for their appearance.

When you investigate the range of mammalian species, there is a remarkable association between baseline heartrate and achievable lifespan. From mouse to whale, the smaller the animal, the faster its heart beats and the shorter it lives for. It is as though mammals have a set number of heartbeats per lifetime, with a mean of 730 million across species.[94]

It turns out that present-day humans buck the trend in a big way, almost tripling our 'natural' longevity. To achieve a lifespan of 80 years, a conforming mammal would only be allowed one heartbeat every 20 minutes, far fewer than the 1,400 times a human heart typically beats during this period. Through our ingenuity and scientific accomplishment, we have converted ourselves into a physiological perturbation. We have become an anomaly of Nature.

Consequently, the healthcare burden over the last century has altered incomparably, with age-related diseases attracting greater research budgets the world over. The US National Institute on Aging, for example, increased its annual budget by about $500 million each year between 2016 and 2021 from $1.5 billion to $4 billion per year.[95]

This shift serves to highlight how human biologists and doctors are steadily served a constantly altering array of fresh challenges. As one problem is solved, another arrives in its wake. Some appear abruptly, such as COVID-19, and others, such as those predisposed by ageing, are more insidious. Maybe some problems are so well hidden they are currently being missed altogether.

A reptilian tale of old

Chameleons are Old World lizards found principally in Africa. Mysticised for such charming attributes as the ability to change their colour spontaneously, lasso their prey lingually and orient their eyes independently, chameleons have long captured the imagination of humans.

In African Bantu mythology, the chameleon is considered the messenger of immortality. However, when tasked with delivering this message of eternal life to humankind by the Creator, the slow chameleon was beaten by the fast-moving lizard, who brought mortality to humans instead. Inspired by such folklore linking chameleons with life and death, Malawi-born Jack Mapanje published his collection of poetry, entitled *Of Chameleons and Gods*, in 1981. He was subsequently imprisoned in totalitarian Malawi for four years, and his poetry was banned there owing to its politically dissenting theme. It is testament to the cryptic nature of his poetry that it was approved for publication by the censorship board in the first place. The chameleon, after all, hides masterly in plain sight.[96]

The Common Chameleon *Chamaeleo chamaeleon* is a species native to the Mediterranean basin and the only wild chameleon species in Europe; the Ria Formosa National Park in southern Portugal is a hotspot for these elusive reptiles. So, in April 2023, while holidaying in this area, my family and I went looking for one.

At the entrance to a coastal pine woodland at Lagoa de Aldeia Nova, signs highlighting the local presence of the famed chameleon were a positive start. Not deterred by the fact that most sightings occur in the autumn when females descend the trees to lay their eggs, our foolhardy optimism led us to think 'maybe, just maybe'. We strolled among the aromatic Maritime Pines *Pinus pinaster* with necks cocked and eyes combing the elevated branches. Mammalian seekers keenly pursued reptilian hiders. Our hopefulness, however, could not have been more misplaced, for there was to be no sighting on that occasion.

This outcome is no surprise when one considers how exquisitely evolved the chameleon is in its natural habitat, and how inherently haughty the human is in his or her exploration.

Equipped as it is with layers of specialised structures known as iridophores in its skin, the Common Chameleon is able to colour-shift to aid its survival. These structures comprise tiny crystals of a molecule named guanine. Under the instruction of hormones and neurotransmitters, the shape and position of these guanine nanocrystals in the chameleon's skin can be controlled. And this determines how light of different colours is scattered, resulting in the luring iridescence of these animals.[97]

While this helps these arboreal reptiles to blend in among the treetops, thereby avoiding the attentions of hungry predators (and holidaying humans), this ability is not just about camouflage. Courtship display, communication and control of internal temperature are other vital roles. Striking the optimal balance between these essential components of survival is the chameleon's evolutionary forte.[98]

Remarkably, the guanine in the skin of chameleons is the same guanine belonging to a quartet of bases that, when attached to a sugar-phosphate backbone, forms the fundamental unit of DNA. As I explored in Chapter 2, Miescher discovered what ultimately turned out to be DNA in 1869. It is difficult to think of a chemical entity more synonymous with life. DNA is the mechanism for building and transmitting life, and yet one of its basic components has found an alternative existence in the reptilian skin. The versatility of life's building blocks is staggering, and it is observations like this that make Nature so fascinating.[99]

This is even more mind-boggling when one considers evidence for the extra-terrestrial origin of several organic molecules such as guanine. Could guanine have arrived on this planet via meteorites before occupying the biological niches we observe today? Perhaps we will never know for sure.[100]

'Guanine' was coined in the nineteenth century, derived from the Spanish word 'guano', which itself stems from the Quechuan word 'wanu'. The Quechua people reside predominantly in the Peruvian Andes, and wanu refers to the dried excrement of fish-eating sea birds deposited along the Pacific coastline. Valuable as a fertiliser owing to its high content of urates, oxalates and phosphates, guano is also a rich source of guanine. Indeed, it was in a sample of guano that guanine was first identified.[101]

As three seemingly rather disparate entities, guano, DNA and the chameleon are unified by the discovery of guanine. Of course, this is a connection that existed way before our species came along. Waste and nourishment; death and survival; alien and terrestrial; hidden and seen. Guanine represents the dichotomies of life.

A marvel of modern therapy

Finding a common relationship between these opposites is a helpful metaphor when it comes to thinking about human disease. While it is often the case that a disease results from the acquisition of a harmful external entity (e.g., the COVID-19 virus), disease can also result from the loss of something essential. Dis-ease is the loss of ease.

No healthcare philosophy understands this better than traditional Chinese medicine. Dating as far back as 8,000 years ago, this system for understanding human health and illness devotes itself to the yin and the yang. These opposing concepts carry a variety of meanings depending on the context, but generally the yin relates to depletion and the yang implies repletion. Cold, dark, weak and passive are the yin, while hot, light, strong and active are the yang.[102]

To explore this general concept further, let's look at the sensational advances in treating a severe childhood neurological disease called spinal muscular atrophy.

Affecting one in every 10,000 children, the most severe form of this disease leads to progressive muscle weakness in babies before the age of six months. The problem is in the nerves that transmit electrical signals from the spinal cord to the muscles. And it's the muscles responsible for breathing, limb strength and trunk stability that are involved. Tragically, the most severely affected individuals are never strong enough to sit, crawl or walk, and they only survive beyond the age of two years with the aid of a continuous breathing machine.[103]

The genetic cause of the disease is well known. But to understand this, it is necessary to first outline some key biological principles.

The instructions for every protein needed for normal cellular function are written in your DNA. As Watson and Crick discovered in 1953, the structure of DNA is a double helix, whereby one string of bases coils up with another string of complementary bases. Specific regions of your DNA, known as genes, contain the instructions for specific proteins.[104]

As a rule, you inherit two copies of each gene, one from each of your biological parents. However, some genes occasionally mutate, and these mistakes can affect the instructions for protein-making in a huge way. The error may only be small at the DNA level, such as the substitution of one base for another, but this can alter which amino acid (the building blocks of proteins) is chosen by the cell when it comes to making the protein. In many cases, these errors significantly affect the function of the produced protein, often making it non-functional or causing its own premature destruction by the cell.[105]

In the case of spinal muscular atrophy, the mutated gene is called survival motor neuron 1, or *SMN1* for short. Identified in 1995, we know that mutations in both copies of the *SMN1* gene (one copy from each parent) are required to cause this severely life-limiting disease.[106]

It necessarily follows that individuals with just one mutated *SMN1* gene do *not* get spinal muscular atrophy. In fact, approximately 1 in 50 of us are carriers of a single defective *SMN1* gene without ever knowing. Typically, it takes two unaffected parents, each carrying a defective *SMN1* gene, to bear offspring with spinal muscular atrophy. In this scenario, there is a 25% chance of an affected baby. In rarer circumstances, the mutation appears *de novo* (from new) during the process of making an egg or a sperm.

If these are unfamiliar concepts to you, then I appreciate how genetics can be a little confusing at first glance. The take-home message at this stage is as follows. For a disease such as spinal muscular atrophy to manifest, both copies of the critical gene need to be mutated. But why is this double loss of the *SMN1* gene so devastating?

The first thing to appreciate is that the *SMN1* gene is just the written instruction for cells to make SMN protein. This involves a two-step process, utilising two areas of the nerve cell called the nucleus and the cytoplasm. If you imagine a cell as a boiled egg, then the nucleus is the yolk and the cytoplasm is the surrounding white part.[107]

In the nucleus, the DNA within the *SMN1* gene is read to make a closely related molecule called messenger RNA (ribonucleic acid). This messenger RNA then moves from the nucleus into the cytoplasm to guide protein manufacture. It does this by providing the template for the assembly of amino acids, which are the building blocks of all proteins. Together, both transcription (conversion of DNA to messenger RNA) and translation (conversion of messenger RNA to protein) form the universal language of life.[108]

The devastating manifestations of spinal muscular atrophy at such a young age make it clear that the SMN protein plays a critical role within cells. Its functions are diverse, involved in cellular house-keeping and energy production, as well as key pathways for synthesis and degradation. While the neurological manifestations of spinal muscular atrophy make it obvious that nerve cells depend on SMN protein, in truth it is made by all cell types in the body.[109]

Hinting further at its vital function in humans, there exists a back-up copy of the *SMN1* gene, called, rather uninventively, *SMN2*. This spare version is almost identical to the main gene, but it differs in one important way. The replacement of just a handful of bases is enough to send most of the messenger RNA created from the *SMN2* gene for immediate destruction by the cell. As a result, the *SMN2* gene only produces about a tenth of the fully functioning SMN protein that the *SMN1* gene can. Nevertheless, this rather wasteful behaviour by the cell has proven to be somewhat of a saviour when it comes to treating spinal muscular atrophy. Let me explain how.[110]

We have already seen how mutations in both copies of the *SMN**1*** gene are necessary to cause spinal muscular atrophy. However, the severity of the condition is correlated instead with the number of *SMN**2*** gene copies an individual has.[111]

It turns out that the region of DNA that contains both these genes is particularly prone to random duplications. Therefore, if an affected individual has just the ordinary two copies of the *SMN2* gene, they will suffer from the most severe manifestations of the disease. However, if there are four or five copies of the *SMN2* gene, then this leads to much milder disease, often with onset in adulthood and sometimes with a normal lifespan.

The extra *SMN2* gene copies mean that, despite the gene's relative inefficiency at producing SMN protein, enough fully functioning protein can be made by cells to delay disease. Each copy of the *SMN2* gene does its bit, meaning that the total rescuing dose of SMN protein is higher, and a state of ease is maintained for much longer.

It quickly occurred to scientists that Nature's remedy could be harnessed artificially. If there were a way of boosting the efficiency of the *SMN2* gene in the most severely affected individuals (those with just two copies of the *SMN2* gene), then their disease could be converted into a much milder form.

Indeed, that thought process culminated in one of the greatest therapeutic success stories in recent times. In December 2016, a new drug for spinal muscular atrophy was approved by the Food and Drug Administration in the USA. This drug is injected into the fluid that bathes the spinal cord, whereby it gains the best access to the nerve cells most affected by the disease. It works to increase the efficiency of the *SMN2* gene by shielding its corresponding messenger RNA from intracellular breakdown. Consequently, more of the vital SMN protein is made.[112]

This approach was so successful that the main drug trial had to be terminated early. During the trial lasting 13 months, just over a half of the babies receiving the drug had a positive response in terms of standard motor milestones, such as sitting, rolling over and crawling. This was compared to *no* babies achieving the same positive milestone response in the placebo group. It would have simply been unethical to continue testing this clearly effective drug against a group of babies purposefully excluded from receiving it.[113]

In just 21 years from the first description of the causative gene, the outlook for a baby born with the most severe form of spinal muscular atrophy has been revolutionised. This is a disease caused not by the addition of something harmful but by the loss of something integral. By observing the biological processes underpinning the disease in great detail, it was discovered that Nature had already come up with the answer. All it needed was some encouragement. For affected individuals and their families, this pharmacological boost can change their lives immeasurably. For all of us, this story is a powerful message of loss and recovery at the hand of natural processes.

Seeing beyond blind faith

Modern doctors hold a problem-solving skillset that is honed over decades of tuition and experience, built to cope with uncertain, gap-ridden data. At the heart of everything is the patient. Doctors must first listen to the patient's history, paying particular attention to the chronology and pace of symptoms. Following this, doctors examine the patient, formulate a working diagnosis and plot an investigative pathway. Treatment is initiated when the chance of benefit outweighs the risk of harm. Published clinical trials equip doctors with the best evidence for and against specific treatments in specific diseases.[114]

Sometimes treatment needs to be urgent, such as in ischaemic stroke, where a blood clot suddenly starves a part of the brain of blood. In this common condition, 24/7 emergency stroke pathways have necessarily evolved towards providing a focused history, a refined examination, a CT head scan and a go/no-go decision on potentially life-saving intervention within tens of minutes.[115]

This general approach appears logical, maybe even obvious, to a modern audience. But this was not the case at its earliest origins approximately 2,500 years ago. Hippocrates of Kos, widely considered the father of modern medicine, said of the medical approach in the fifth century BCE: 'The art consists of three things: the disease, the patient and the physician. The physician is the servant of the art, and the patient must combat the disease along with the physician.' This was a new way of thinking at the time.[116]

Prior to Hippocrates, disease was believed to emanate from a supernatural world, often as a punishing handout from the gods. This was at a time when knowledge of biology was effectively non-existent. The anatomy of the human body was not known owing to laws forbidding human cadaveric dissection. The cellular underpinnings of bodily function were hidden in an invisible microscopic world. Bacteria and viruses laid undiscovered despite the havoc they wreaked in frequent plagues. The genetic code of life operated incognito. And without a biological theory to explain human health, no viable framework for disease could develop. Instead, healing was the remit of magic, religion, incantation and prayer.

Asclepius lived as a healer in Greece at least 300 years before Hippocrates. Over the subsequent centuries, he became immortalised in Greek mythology as the god of medicine and healing. First mentioned in Homer's epic poem *The Iliad* in the late eighth century BCE, Asclepius the man began his documented transformation into Asclepius the myth.[117]

As later versions of the myth portray, Asclepius was the offspring of the Greek god Apollo and a mortal woman. However, soon after conception, Apollo became jealous of the affections Asclepius' mother paid to a mortal man. So Apollo burnt her and ripped the unborn Asclepius from her womb. Apollo gave his son to the wise centaur

Chiron and told him to teach Asclepius all there was to know about medicine and healing.[118]

Asclepius went on to accomplish miracle cures for the breadth of human ailments, including one for death itself. When Zeus (King of the gods and Asclepius' grandfather) heard of these human resurrections, he felt betrayed and outraged. In response, Zeus struck Asclepius with a thunderbolt, simultaneously releasing him from a mortal, hominin life and bestowing upon him an immortal, divine one.

This story typifies the human-like, emotionally charged relationships that not only justified the process of deification, but also rendered it relatable to the Ancient Greeks. Given that Asclepius defeated death and riled the almighty Zeus into action, his healing powers must have been prodigious. When you found truth in this story, you were rewarded with the faith that Asclepius served to heal you. In the god's temples that became widespread in Ancient Greece, known as the Asclepieia, the reverent and the revered belonged to a common understanding, reinforcing Asclepius' divine status. By killing the mortal Asclepius, the all-powerful Zeus had permitted the elevation of Asclepius to Olympian immortality, while preserving his Earthly duty.[119]

Asclepius had become the eternal healer. In transacting his power over health and disease on Earth, he summoned help from the animal kingdom. And the snake was his most famed companion. It is said he once cured the muteness of a young girl by sending a snake to scare her into speech. Even today, the symbolic influence of the snake in medicine continues. The World Health Organization, the American Medical Association and the British Medical Association all contain the Rod of Asclepius in their logos, which depicts a snake wrapped around a rod.[120]

And it is not difficult to see why the snake became the perfect associate for Asclepius. Not only does a snake regularly shed its dead skin to reveal a fresh lining, but the venomous ones also possess the anti-venom to their own toxin. These are powerful symbols of the snake's authority over life and death, and, by extension, over health and disease. But while the lure of such symbolism remains difficult to relinquish fully, the need for a more rational approach soon became necessary for many Ancient Greek physicians.

Hippocrates is credited with the creation of the first plausible biological theory of health and disease. He based it on the four humours, namely black bile, yellow bile, phlegm and blood. Analogous to the four natural elements, water, earth, wind and fire, which were believed at that time to be the balanced components of the universe, Hippocrates proposed that imbalance of the four humours led to human disease.[121]

He was utterly wrong, of course. But that is beside the point. Importantly, the very idea that human disease had a rational basis laid the foundations for how medicine is practised today. Close observation of the symptoms, rational formulation of a diagnosis and judicious initiation of therapy all found their basis in Hippocrates. Indeed, he invented these three medical terms, which remain in wide use today. Modern theory of disease may have dramatically altered since Hippocrates, but the stringency of his clinical method very much remains.[122]

First comes tolerance of the unknown

One man of antiquity who brought us a step closer to the molecular basis of disease that dominates today was Asclepiades. Practising as a Greek physician in the Roman Republic at the turn of the first century BCE, Asclepiades shunned the four humours theory. Instead, he took inspiration from the Atomists and considered the human body to be made up of atoms separated by space. He thought of this space as a series of pores, which became either too wide or too narrow in disease. He imagined the atoms existing in varying combinations throughout the body, constantly moving, colliding, splitting and reforming.[123]

This theory provided Asclepiades with a way of explaining common symptoms. For example, fever resulted from the friction induced by bodily fluids as they passed through constricted pores. And while dryness of the mouth and constipation happened when the pores were blocked, perspiration and diarrhoea developed when the pores were wide open.[124]

Asclepiades paved the way for the Methodist School, arguably the most influential medical school in the Roman Imperial era. Set up by one of Asclepiades' students, the Methodist School attempted to amalgamate two popular but polarised philosophies of the time: one based on dogma and the other on empiricism. The Dogmatists pursued theory and logic above all else, whereas the Empiricists trusted observation and generalisation in the real world. The Methodists attempted to find a middle ground.[125]

They formalised Asclepiades' atom-pore theory, bringing in terms such as *status strictus* (blocked pores), *status laxus* (gaping pores) and *status mixtus* (mixture of the two states). In addition, the Methodists characterised illnesses as either acute or chronic, while acknowledging that there existed a spectrum between the two. At one end, acute illnesses were of rapid onset, greater intensity and shorter duration, while chronic diseases grumbled along at lower intensities for longer periods of time. This is a familiar concept to modern doctors, but it was a novel approach in the early Roman Republic. Many still believed in the 'critical day' concept, which relied on solar and lunar cycles, mixed with a considerable dose of blind faith and superstition, to decide the patient's outcome.[126]

The Methodists' call for rationality and balance is epitomised in this description by the first-century Roman encyclopaedist Celsus in his only surviving work, *De Medicina*: 'Some of the Methodists of our own age ... assert ... that sometimes the excretions of sick people are too small, sometimes too large; and sometimes one particular excretion is deficient, while another is excessive. That these kinds of distempers are sometimes acute, and sometimes chronic; sometimes increasing, sometimes at a stand, and sometimes abating.'[127]

Having the confidence to leave one's observations temporarily unexplained was a refreshing change. This approach required a tolerance of uncertainty, which takes us back to the main idea in Chapter 2, whereby the unknown knowns are what bound us. My use of the word 'bound', with its dual reference to something that both limits and leaps, is not accidental. When we acknowledge the unexplained components of our worldly observations, it not only creates an obvious boundary, but it also enables us to make strides beyond. From Asclepius the god of medicine to Asclepiades the practical physician, ignorance had become acceptable, fashionable and even desirable. Ignorance, as the food for undiscovered knowledge, had become bliss.

Stories dominate our lives

Collectively, our propensity to tell tales is stronger than our instinct to discover the rational truth. From an evolutionary perspective, stories are vital for the survival of our species. Stories empower us, bind us together and inspire us. As a social species, gossip, legend and folklore have facilitated our linguistic and cognitive development. This has permitted high-density living, affording the protection, specialisation and cooperation characteristic of successful human civilisations. Our ability to imagine the fictional is one of evolution's greatest accolades.[128]

But fiction at the glaring expense of logic can be a problem. A story in unwavering defiance of rational fact can be an even bigger problem. It is a problem because powerful stories take hold of our imaginative pathways, shutting the door to alternatives. After all, these pathways have been naturally selected over millennia for their ability to latch onto the fantastical. They aid our survival, because they unite individuals amid a common understanding of a seemingly random and uncertain world. This is what makes an unbending, fact-defying story so hazardous. People believe, people follow – and people submit.

Meanwhile, flexible narratives that are permitted to mutate, branch, grow and recede are a different story. Hippocrates devised his four humours narrative; Asclepiades concocted his atom-pore idea; and Miescher acted on his nuclear hunch. All these theories first required human imagination, which then set a more focused agenda. None of these theories was spot on (some far from it), but none snubbed a subsequent upgrade or two either.

In Chapter 1, I described the classical view of brain function, whereby the left hemisphere is viewed as dominant and the right hemisphere as non-dominant. This stems from the fact that the left side controls speech and handedness in most people and is therefore the most direct way we express ourselves to the outside world. As precision is important to the left hemisphere, its functional range is relatively inflexible, coping best in familiar, well-categorised contexts. In contrast, the right hemisphere is often considered the left's inarticulate and gauche sidekick.[129]

However, this classical split between the two hemispheres has been eloquently challenged. The psychiatrist and neuroscientist Iain McGilchrist wrote in 2009 that the left side of the brain is instead the rebellious emissary to its right-sided master. In this model, the gab and grab approach of the left hemisphere is subservient to the more holistic and flexible world overview exclusively housed in the right hemisphere. And in neglecting the real supremacy of right-sided attributes, such as metaphor, music and humour, modern society has fallen victim to the left's ill-considered whims. 'In short,' says McGilchrist, 'the left hemisphere takes a local short-term view, whereas the right hemisphere sees the bigger picture.'[130]

While humans remain at the helm, our approach to healthcare is beholden to the distinct natural qualities of both halves of our brains. We must balance our right-sided imaginary tendencies with our left-sided pragmatic duty, embracing continuous dialogue between the two. In the same way that the Methodists attempted to merge left-sided dogma with right-sided empiricism, we should be striving for the perfect combination of informed ignorance and rational refinement. In doing so, we might just benefit from the best of both brainy hemi-worlds.

Reclaiming what's been lost

In these opening three chapters, I have introduced what I consider to be the key concepts propping up this book's core message. However we make sense of the natural world, it affects what we know and how we feel about it. And in conveying this process of thought, Nature has customarily been my chosen metaphor. But it does not matter whether the focus of enquiry stems from a scientific, artistic, conceptual, spiritual, philosophical, literary or historical bent, for a bit of each is invariably present. In the same way that humans are intricately embedded in the natural world, so too are all these perspectives in any real-world problem we face. As individuals with distinct personalities and experiences, we each take our fancy to one or two of these domains, only overlooking the others to our detriment. I hasten to add that this list of ways to approach a problem is in no way comprehensive. Undoubtedly, the complete list is as boundless as the issues it serves to tackle.

Nature asserts its rule wherever we probe. Every living being, including ourselves, is a node within a complex network, each with inputs, outputs and internal circuits. Like it or not, we are at the mercy of these natural interactions. And as a neurologist I am regularly confronted with impairments in human brain and nerve function. Whether the cause of dysfunction is a structural, developmental, infective or immune-based one, disease results from a disruption of cellular harmony. There may be too much of one thing, causing toxicity, or too little of another, leading to deficiency. Add to the mix the biological influences of natural time cycles, the linearity of ageing and the incomprehensibly protracted pre-human era, and the complexity builds somewhat. Every interaction at the atomic, cellular, individual or ecosystem level is moulded by the environment in which it acts. Consequently, these interactions have evolved to work best in familiar settings when presented with familiar stimuli. Deviations are unsettling and, when sustained, result in distress.

The stressful vessel model conceptualises both the obstinacy of stressors and the dynamism of effective coping strategies. But as a model, it is a replica of reality and must be constantly moulded, then remoulded, to maintain its fitness for purpose. It must contend with the façade of knowledge, which stands taller than ever in our modern, more connected world. Unknowns, uncertainties and inaccuracies are part and parcel of a robust scientific method, requiring sound statistical supervision. In that way, progress can at least be a realistic objective.

While theories are satisfyingly eloquent stories, models are stories shaped by observation. Narratives bind science together, just as they have bound extinct and extant human communities over many hundreds of millennia. Even over a period as short as the last 2.5 millennia, our approach to human disease has necessarily adapted to the varying strains placed upon human health. But to enhance the greatest power it can offer, that is *to predict the future*, the healthcare approach is necessarily shaped by the past. We guard the best parts and disregard the rest. The methodical nature of Hippocrates' clinical approach remains almost untouched to this day, while his four humours theory has been retired. The myth of Asclepius is nowadays seen for the elaborate story it really is, but influential health organisations today still hold onto the appealing symbolism of Asclepius' healing snake.

And from one reptile to another, it is perhaps the chameleon that best symbolises the obscurity of human health. As a spiritual representation of immortality in many African cultures, the chameleon's inconspicuousness and inconstancy are the precise aspects of human disease that most bedevil us. The nature of the interplay between patient, healthcare professional and disease is in constant and veiled flux. Remaining alert to this fact is vital for those tasked with setting and implementing the modern healthcare agenda.

My central argument is that the evolutionary make-up of our species, just like any other, embeds us in the natural world. For me, it is the one narrative upon which all else hangs. Hippocrates recognised 'benevolent Nature' in the therapies he recommended to patients. This was reflected in the diets, herbal remedies and exercise he relied upon. But perhaps, in the midst of our modern medications, technologies, analyses and cities, we have lost sight of Nature's benevolence. While the natural world remains as communicative as ever, its restorative powers have been steadily lost in translation. In the remainder of this book, I hope to convince you that our reclamation of this loss is long overdue.

And so I resume, in Part II, by exploring the closest and most dynamic relationship of them all: the one between planet Earth and life itself.

PART II

THE ROLE OF THE EARTH FROM LIFE'S ORIGIN TO THE MODERN ERA

Dark Marbled Carpet

CHAPTER 4

The Climateric

The first raindrop of a storm lands imperceptibly

If the seed is to the tree what the egg is to the chicken, an obvious question arises: which came first? This type of paradox is as old as it is devious. If the tree came first, from what did it germinate? But if the seed came first, what produced it in the first place?

The automatic response is to consider the seed and the tree as separate entities, as that is how they are presented to us. Indeed, the dominance of our categorical left hemispheres in processing speech imposes this separation. However, it is this duality that creates the dilemma. Instead, think of the seed and the tree as two parts of the same biological entity, and you realise the very supposition that one of them *must* have come first is fundamentally flawed. They both appeared simultaneously!

But surely the chances of that happening are impossibly tiny, you may be thinking. Not if you consider the tree and the seed as intrinsically codependent, without either of which the whole would never have existed at all. You either have both, or you have neither.

Let me illustrate with a thought experiment. Imagine the simplest unit of life that can reproduce. I will call it O. When O reproduces, you end up with two Os: O O. Time passes, and many Os are spawned perfectly. However, every now and then, a modification to the reproduction process occurs. Some of these alterations are harmful, and the resultant unit of life, for example C, does not survive long enough to reproduce further. But some changes, for example D, *are* compatible with reproductive life, albeit in a slightly altered form.

In this hypothetical pool of alphabetical life, there are now a mixture of Os and Ds, separately reproducing to make more Os and Ds. Over time, another random variation may appear that produces

P instead of D. This turns out to be compatible with reproductive life too, and now a population of Ps establishes itself. As the complexity increases, it becomes apparent on closer inspection that a **P** does not just instantaneously divide to produce another **P**; it requires a couple of stages. First, the line **I** grows, before the upper loop grows in turn, thereby forming **P**. Subsequently, it isn't a huge leap for the loop-making ability to become duplicated, thereby generating **B**. Over time, these steps become more intricate. The successive generation of the I group, say **J**, differentiates itself further from the successive generation of the B group, say **E**. The Js and the Es are now distinct enough to be seen existing together. But they remain inextricably linked, for each **J** turns into an **E** via a predictable series of steps: J-I-P-B-E. To being the cycle again, E produces a new J.

While this is just the start of a potentially endless process, I hope it seems intuitive that the J group represents the seed and the E group symbolises the tree. Yet neither could claim to have come first; they appeared in tandem as a result of this divergent, iterative process.

This can be a challenging concept to grasp and may come across as unnecessarily abstract, maybe even pointless. Why am I talking about replicating and transforming letters of the alphabet? In short, it is my attempt at explaining the shifting stability of biological states by encapsulating the idea within more familiar terms. Ultimately, this chapter is about the interactions that drive this endlessly dynamic process.

Nothing operates in isolation

To elaborate further, I want to introduce some numbers into the thought experiment. Let's assume the numbers represent a parallel thread of life, whose origin is represented by 0. Its similarity to **O** is no accident, for both are governed by the same intrinsic biological rules. Indeed, both had the same ancestral origin. Most of the time, 0 reproduces perfectly, but as we have come to expect among these characterful creations, replication mistakes can occur. Over time, populations of 6s, 9s and 8s begin to emerge among the 0s.

At this stage, the letters and the numbers lead separate reproductive lives, lacking the ability or need to interact. However, at some point a modification among the 8s leads to the emergence of 3, which happens to interact favourably with a specific component of the parallel thread of life, namely E. Their three prongs align, E≡3, and this somehow promotes the reproductive lives of both populations, the Es and the 3s.

Perhaps this symbolises the rudimentary stage of something familiar, such as the dispersion of a tree's seed by an animal attracted to the tree's sugary fruit or the symbiotic relationship between Common Alder trees and *Frankia alni* bacteria from this book's introduction. But I want to get away from this being an interaction that *must* be understood at the human level. This interaction can work at multiple unseen levels, including molecular, cellular, anatomical and physiological ones. The key point is that these letters and numbers represent reproductive biological states that stabilise themselves for the most part, while retaining the capacity to shift and interact when needed (Figure 4).

Although I acknowledge that the term 'reproductive biological state' carries ambiguity, it also commands a breadth of application that is useful. Humans reproduce; Veeries reproduce; alders reproduce; bacteria reproduce. Even skin cells and immune cells reproduce. And

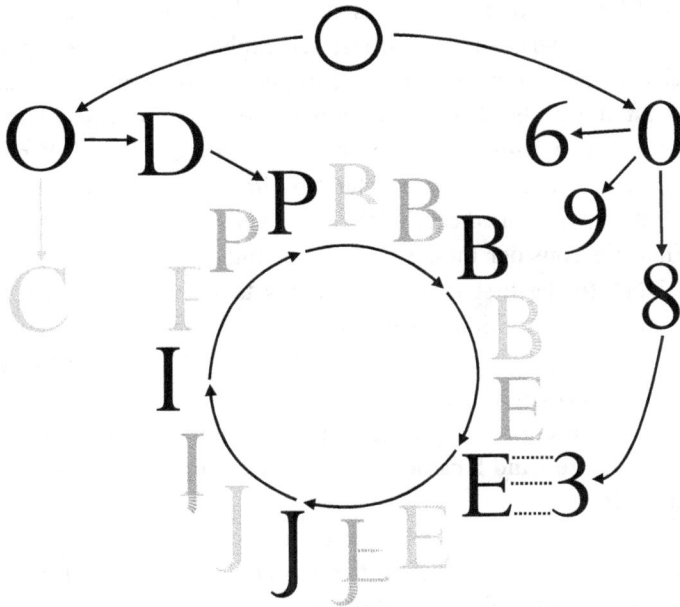

Figure 4. Hypothetical cycle of reproductive life: each character represents a biological state that may stabilise, shift or interact.

sperm and egg cells are unique in that they permit sexual reproduction of their corresponding species.[131]

Undoubtedly, single cell reproduction is the simplest scenario
we are drawn to. But single cells did not materialise out of thin air.
They are made up of many complex components, whose interactions
throughout the course of evolution have governed the assembly of
the cells we observe today. If we strip a cell back to its origins, what
is known about the first reproductive unit? Can biological molecules
even self-replicate in the absence of a cell's finely tuned reproductive
machinery?

Going right back to the beginning

Briefly cast your mind back to Chapter 3, where I explained how
spinal muscular atrophy has become a treatable disease. In order to
achieve this, scientists required detailed knowledge about DNA, RNA,
proteins and their regenerative interactions. These are components
that generally operate *within* the confines of a cell. But what happened
before the first cell?

The RNA world theory states that RNA uniquely satisfies the
fundamental criteria of the earliest self-replicator, way before the
first cell ever existed. In general, a self-replicator must fulfil two main
tasks. First, it must be able to store information about its own identity
that can act as a template for the next generation. Second, it must
be capable of triggering and accelerating its own reproduction in a
consistent way (so-called catalysis).[132]

When we consider biological systems today, we would typically
assign DNA to the first task and proteins to the second. However,
since DNA is just modified RNA, and RNA is the code that guides
protein assembly, it is more than feasible that RNA once performed
both roles on its own, way before DNA and proteins even existed. This
is like the **P** from our thought experiment going about its reproductive duties before **I** and **B** came about. Let's examine this possibility in
a bit more detail.

RNA certainly shares DNA's credentials as an efficient store of
information. Its fundamental unit of storage, the base, comprises a
restricted cast of four players, just like DNA. In Chapter 3, I explained
that one of these bases is guanine, which crops up in chameleon skin
and seabird poo to boot. When it comes to DNA and RNA, guanine

always pairs with its complementary base, cytosine. And these two bases are held together by three chemical bonds, resembling the E≡3 interaction from the thought experiment just mentioned. Crucially, this differs from the two chemical bonds that connect the other two bases.[133]

What this strict exclusivity brings is fidelity. Consider the double-helix structure of DNA, which was confirmed by Watson and Crick in 1953. The two strands of this information powerhouse are complementary, meaning that wherever there is a guanine base on one strand, there is a cytosine base on the other.[134]

And this unbreakable fidelity bred notable success. If all the DNA double helices within a single human cell were fully uncoiled and laid end to end, they would measure just over 2 m. Extending this to the 3 trillion nucleated cells making up each human body, the total length of DNA in a single person would cover over 20 return trips to the Sun![135]

While this is clear evidence of DNA's role as life's universal microchip today, this represents far more information than the earliest self-replicator would have been required to carry. And there is a natural tendency to overcomplicate a biological question like this, given we are constrained to approach it from an advanced position in the evolutionary saga.

In an attempt to avoid this trap, experts have simulated the more primitive conditions and chemistry that were predominant at life's origin. Although such an ancient issue will continue to resist definitive proof, scientists in this field have proposed several promising models. None exists without its critics, but it does seem that given the optimal combination of temperatures, water content, ultraviolet light, sugars and minerals, the very first RNA molecule could have formed by 3.5 billion years ago. We are unlikely to ever know what happened for sure.[136]

This prompts us to consider the second criterion that the very first replicator must have satisfied: catalysis. For RNA to have achieved the greatest biological milestone, *the very inception of life*, then it must have developed the ability to trigger and accelerate the reactions involved in its own reproduction.[137]

While RNA-driven catalysis was initially thought to rely on very long chains that would have been unstable in that early environment, there are now numerous examples of much smaller RNA units demonstrating such activity. Basic actions, such as chopping and joining, would have been instrumental in the reproduction process. Undoubtedly, the rates of catalytic enhancement would have been modest in those early reactions compared with modern standards. But even those RNA molecules capable of the smallest enhancements would have favoured their own self-replication above that of the crowd.[138]

Ultimately, a primitive form of RNA may have been the molecule that satisfied both essential criteria of the earliest self-replicator, opening the door for Darwinian evolution as we understand it today.

Those capable of adapting endure

There is perhaps no greater endorsement of evolution's prowess than the Texan winter adaptations of the 'American chameleon'. Surprisingly, this animal is *not* a chameleon, for there are no naturally occurring wild chameleons in the Americas. It is instead a lizard, known as the Green Anole *Anolis carolinensis*. In Chapter 3, I mentioned that the lizard is considered the deliverer of human mortality in African Bantu mythology, triumphing over the slow chameleon and thereby preventing the chameleon's intended gift of human immortality. This Green Anole lizard, therefore, offers the perfect reality check after our foray into an imaginary alphanumeric world and the remote fortunes of early RNA.

In the winter of 2013/14, a weakened polar vortex permitted the escape of cold Arctic air across North America. This led to atypical sub-zero lows (<0 degrees Celsius; <32 degrees Fahrenheit) in relatively warm southern states such as Texas and Florida, breaking winter temperature records that had stood for over 100 years.[139]

This cold snap was an extreme weather event, providing a rare opportunity to learn about the adaptability of the Green Anole lizard over a single season. Native to the south-eastern regions of the USA, this lizard demonstrates variable tolerance to plummeting winter temperatures depending on where it lives. Those commonly exposed to colder winters in Oklahoma, for example, remain active at lower temperatures than their cousins on the Mexican border. As luck would have it, this spectrum of temperature tolerance had been well characterised in the summer of 2013, providing the perfect baseline to evaluate the effects of the widespread winter freeze that followed.[140]

When the lizard populations were reassessed in the spring and summer of 2014, the lizards living in the warmest locations had aligned their behaviour with those living in the coldest. They, too, were now able to operate at lower temperatures, eradicating the discrepancies found between northern and southern populations just the year before. The winter storm had pushed the comparatively cold-intolerant lizards of the south to the edge of survival, and only those individuals capable of adapting to the strain endured. This was survival of the fittest playing out over just a few months.[141]

Even more remarkable than the observation itself is its putative mechanism. But before I explain, a quick recap of some basic cellular biology is warranted.

DNA is the universal genetic code of life, and by employing an RNA intermediary, it indirectly encodes the making of proteins. Each gene in your DNA is responsible for one or more proteins that then go on and influence the way the cell functions. This might be to help shuttle components around the cell, provide stability to the cell's architecture or aid in modifying the protein production line. While DNA *instructs*, proteins *do*.[142]

As it would be disadvantageous to have all your 20,000+ genes being converted into proteins all at once, genes in your DNA can be turned on or off as necessary. There are two main ways the cell achieves this. The first is by attaching a simple chemical group, made up of one carbon atom and three hydrogen atoms, to DNA bases. If enough of these groups (so-called methyl groups) are added to dedicated promotor regions of DNA, the gene is silenced, and the corresponding protein will not be made.[143]

The other main method at the cell's disposal involves altering the way DNA is packaged up inside the nucleus. DNA carries a negative charge, and it is usually tightly bound to positively charged proteins called histones. These histones can be chemically modified to either tighten or loosen their grip on DNA. And as their grip loosens, DNA is made available for converting its message into protein.

As you can probably appreciate, this biological system is complex, requiring close regulation to maintain cellular health. Indeed, imbalances within this process have been shown to occur in certain cancers, as the wrong mix of genes are turned on, and the resultant proteins act to induce uncontrollable cellular division.[144]

With that in mind, let's now return to the plight of the Texan Green Anole lizards. How does this knowledge help to explain their fast adaptability to the extreme cold?

Among the southern population that was the least accustomed to the very cold winter temperatures of 2013/14, there were clear changes in the composition of 'turned on' genes in the lizards sampled. Genes associated with the way the lizard's nerve cells communicate with muscle cells had altered their activity as a direct result of this single seasonal anomaly. This might have made the difference between catching life-sustaining prey, or not. Or retreating to a less hostile position at the right moment, or not.[145]

It can reasonably be hypothesised, therefore, that these alterations helped the individuals possessing them to survive, whereas those that were not genetically predisposed to make such changes succumbed. Importantly, the surviving lizards would go on to transmit a set of genes to the next generation that enhances their collective tolerance to future extreme winters. In this way, the lizard's evolutionary struggle lurches on.

Biology is rarely black and white

Although such a stark environmental anomaly stamps an indelible mark on a lizard's biological resilience, the evolutionary march should more aptly be viewed as a plodding, rather than a lurching, one. Generally, an organism will cope better with a gentler survival challenge spread out over many generations, allowing for a slow genetic drift towards a well-adapted population. In finding a suitably illustrative example, I now turn to our own species. More specifically, I turn to the colour of our skin.

'Toubab, toubab,' echoed a chorus of children, accompanied by pointing fingers and wild laughter, as I jogged alongside the Atlantic coastline in Saint-Louis, Senegal. I smiled back, waved and continued on, not really sure if their chanting was meant as a greeting or a slur. It was my first week in the city during my medical school elective in June 2009.

Despite brushing up on the remnants of my high-school French in preparation for the six-week trip, I could not recall any word or phrase that resembled 'toubab'. It was only when I returned to my accommodation in a local family's home that I was reliably informed of the word's meaning. In the local language, Wolof, 'toubab' means 'white person'. It remains the only time in my life I have been knowingly singled out, albeit rather unmenacingly, owing to the colour of my skin.

If only everyone subject to any degree of racial prejudice, either presently or historically, could say the same thing. Amitav Ghosh, in his book *The Nutmeg's Curse*, weaves a time-warping narrative of colonialism, exploitation and xenophobia, arguing that these atrocities set the world on a path to natural decimation. Ultimately, this has culminated in the climate and biodiversity crises we face today. To me, it is heart-breaking that a direct product of natural adaptation, *the way we look*, should have triggered such a catastrophic sequence of destruction and loss. And this compelled me to explore the biological origins of human skin colour in more detail.[146]

Recent evidence shows that the variation in human skin colour that is observed across the globe today is a relatively recent evolutionary addition, originating at some point over the last 30,000 years or so. Prior to this, our species universally possessed a dark skin tone.

But to place this recent introduction of variable skin tone into its appropriate evolutionary context, we need to time-travel several millions of years into our biological past. In doing so, we can trace our ancestors' original transition from hairy light-skinned apes to relatively hairless dark-toned humans. This inherently African tale is about the enduring influence of sunlight and temperature on our ancestors' fertility. The starring cast includes two nutrients that most of us take for granted, namely Vitamin D and folate. And it is a story so fascinating and relevant to our modern self-perception, we should all, in my view, have some awareness of it. So let's begin in the skin itself, shall we?[147]

The human skin photosynthesises. Now before you stop reading owing to the apparent inaccuracy of that statement, I'm not referring to the same photosynthesis that occurs in the green leaves of plants and trees. Since 'photo' means 'light', and 'synthesis' refers to 'making', anything that makes something using light can be said to photosynthesise. What the human skin makes is a precursor of Vitamin D, called Previtamin D. And it is sunlight arriving at the skin that sparks this process.[148] This happens to be a vital initial step in the bodily production of active Vitamin D, which is essential for the health of our bones, muscles, immune systems, hearts, lungs and brains. As you can imagine, Vitamin D is not something we can just do without.[149]

But sunlight had always been in plentiful supply for our ancestors prior to their migration out of Africa 60,000 years ago. In that predictably sunny African environment, during a pivotal period of our evolutionary development, there was thankfully no risk of making insufficient Previtamin D. And therefore there was no risk of Vitamin D deficiency. In contrast, it had been the damaging effects of excess sunlight that had imposed a much greater survival pressure on our ancestors during that time.[150]

Today, we know full well that ultraviolet rays from the sun cause harm. They cause cancers by damaging DNA, destroy key nutrients in the blood and burn sweat glands. But for many animals in equatorial regions, including today's light-skinned chimpanzees, their hair helps to protect them from these dangers. So why didn't our ancestors keep hold of their bodily hair? And why was darker skin tone selected as a suitable alternative? Let me share with you one of the most coherent explanations science can currently offer.[151]

It's all about the environment

About 6 million years ago, the predecessors of our own 300,000-year-old species split from the arm of evolution that went on to give us the hairy, light-skinned chimpanzees of modern day. One can therefore conclude that at some point between 6 million years ago and 300,000 years ago our predecessors exchanged their bodily hair for darker skin. But what drove this adaptation? One fundamental factor was the environment of the time.[152]

Six million years ago, the Earth was emerging from a particularly hot and humid spell in the planet's history, which had produced temperatures many degrees warmer than today. As the Earth slowly cooled it also became steadily drier, leading to the conversion of equatorial tropical rainforests into savannah. This relatively treeless habitat was the scene for a variety of distinctly human attributes to appear, including an upright posture, enlarged brains and bodily hair loss. Just like the seed–tree paradox, it is debated which of these came first. But as the thought experiment at the start of this chapter concluded, one should not discount the probability that such codependent features evolved in parallel. That being said, let's first look at the 'hair loss' bit in more detail.[153]

Hair is a protein plug. What this means is that once protein is deposited in hair, it cannot be retrieved and used by other organs such as the brain. But despite this biochemical dead end, natural selection favoured hair growth early in mammalian development to insulate not only against the cold but also the heat. Like a thermos flask that traps a layer of air to block the transfer of heat in either direction, so too does the thick fur of most mammals. It is the perfect energy cocoon.[154]

It therefore stands to reason that relative nakedness required a particularly strong survival advantage to compensate for the loss of hair. While humans sit at the hairless end of the mammalian spectrum alongside whales and elephants, even chimpanzees are not as hairy as they should be for their size. What this tells us is that the great apes (including our own ancestors) had been losing hair for some time prior to 6 million years ago, perhaps in response to the warmer climate of the era. But what this also implies is that something else uniquely accelerated this trend in humans. Let's examine what the main contenders are.[155]

The building blocks of proteins are amino acids, many of which are not part of the human synthetic repertoire and therefore must be consumed in our diet. Consequently, these essential amino acids constitute a limited and variable resource depending on their dietary availability. Hair happens to contain lots of these amino acids. And so for every gram of protein locked up in hair, it means a gram less for the rest of the body.[156]

One organ that suffers more than most in the face of an inadequate protein supply is the brain. Thus, one theory posits that the suppression of hair growth freed up the necessary ingredients for brain enlargement. This would be a boon for any species in a position to explore new habitats, communicate in social groups and utilise tools.[157]

However, brain cells are a sensitive bunch. In order to function normally, they must be kept within a very narrow temperature range. When too cold, brain proteins slow down and don't work. When too hot, the same proteins get damaged and don't work. Heat stroke occurs when the brain overheats, leading to its characteristic neurological features of confusion, reduced consciousness, imbalance and seizures. Therefore, even if there were evolutionary benefits for an enlarged brain *and* there was sufficient protein to make one, temperature control was vital. Humans needed an improved blood conditioning system.[158]

Sweat and sunscreen

While the loss of bodily hair would have certainly left our ancestors more vulnerable to swings in ambient temperature, it also enhanced the impact of Nature's ready-made coolant: sweat. Great apes, such as chimpanzees, and Old World monkeys, such as macaques, all sweat, positioning this as an ancient thermoregulatory tactic. However, no primate sweats as much as the human. This is partly because hair obstructs the evaporative heat loss upon which sweat relies, but also because our upright posture boosts it. The improved biomechanical efficiency of running on two limbs simply wicks the sweat away, making it incredibly efficient at offloading muscle-generated heat over long distances.[159]

And breathe. It turns out condensing several millions of years of human evolution into a few pages is not easy. Nonetheless, the picture building up so far looks something like this.

Between approximately 6 million and 300,000 years ago, the human prototype was faced with a new habitat, characterised by reduced tree cover and longer spells under the strong equatorial sun. To capitalise on this opportunity, brain power was vital. But there was a problem; larger brains needed lots of protein and tighter temperature control. The suppression of bodily hair growth not only reduced the inevitable dumping of protein associated with a thick coat of fur, but it also increased the efficiency of sweating. Problems solved, you might think. But there was one glaring issue: how to manage the damaging effects of ultraviolet rays on naked skin.

Yet again, Nature had been working on a solution for hundreds of millions of years. The solution was melanin. Displaying a wide variety of forms and functions across all of life's kingdoms, melanin is Nature's homemade sunscreen.[160]

Deriving its name from the Greek word for 'black', the unique structure of this dark pigment, in all its various forms, makes it the best absorber, reflector and dissipator of ultraviolet light in biology. But that's not all. As confirmation of its antiquity, melanin disarms toxins in bacteria, heals wounds in plants, resists microbes in insects, reinforces feathers in birds, blinds predators in cephalopods and resists nuclear attack in fungi. Clearly, wherever organisms meet hostility in their respective environments, melanin has evolved a protective role.[161]

While that hopefully all makes sense, it is worth remembering that the natural selection of seemingly advantageous genes, such as those producing more melanin in our skin, must have been driven by an improved rate of successful reproduction. Without this, it is unlikely that any trait would ever become universal in a species.

In modern medicine, we consider ultraviolet rays as a major cause of skin cancers through DNA mutations. This is something that those with darker skin tones are protected from to a degree. But skin cancers generally come on later in life well after the age most people have children. Therefore, it seems unlikely that this would have been a key driver for melanisation in our ancestors. On the other hand, the destruction of sweat glands by ultraviolet rays would have resulted in overheating and reduced survival, and it is therefore a worthy contender. But it pales into insignificance when compared with the toils of a rather unassuming, water-soluble vitamin, called folate.[162]

Folic acid, as the synthetic version of folate, was worth over $1 billion in global sales in 2024. As adequate folate is essential for a developing foetus, such widespread use today helps prevent serious neurological malformations in babies. The most extreme malformation is anencephaly, which literally means 'no brain'. So how does this fit into our evolutionary past?[163]

Humans can neither make folate for themselves nor store it in large quantities. Therefore, regular intake of folate-rich foods, such as dark green vegetables and tropical fruits, is crucial. But there is a catch. In stark contrast to the photosynthesis of Previtamin D, folate is destroyed by ultraviolet rays as it travels through blood vessels near the skin. Where light giveth, it also taketh away.[164]

Ultimately, then, melanin served as a blockade between light and folate, conserving folate's vital brain-forming power during the human reproductive cycle. This is likely to have created a strong survival pressure for melanisation of our ancestors' skin. And with this, we arrive at the perfect time for a summary.[165]

Big-brained and dark-skinned, we ran our hairless, perspiring bodies out of Africa about 60,000 years ago. Natural selection converged on this combination of attributes, simply and *only* because one tweak after another improved the reproductive success of our evolutionary line when confronted with the slowly changing environment of that period. This was life by a thousand strides.

How farming changed our appearance

Finally, I am in a position to explain the variation in human skin tone observed across the globe today. This part of the tale involves our old friend Vitamin D and its reliance on sunlight for activation, plus a change to our ancestors' cultural livelihood that might surprise you: farming.[166]

As I've already mentioned, genetic evidence suggests that the introduction of gene variants responsible for lighter skin tones in humans most likely occurred at some point in the last 30,000 years. In fact, the most recent evidence increases the likelihood that this characteristic only became widespread in the last 10,000 years or so. Therefore, for approximately 50,000 years after our migration out of Africa, it seems that despite reduced sun exposure away from the equator, a dark skin tone remained the best survival tactic for all the reasons it evolved in the first place.[167]

So what changed? Two considerations are particularly important. First, the end of this period witnessed the melting of large ice sheets, which had covered many regions of North America and Eurasia that are inhabitable today. For tens of thousands of years, these ice sheets would have blocked our ancestors from moving too far away from the equator.[168]

Second, our species made a dramatic cultural change starting around 12,000 years ago. Instead of the hunter-gatherer dominance that had existed to that point, our ancestors began to farm for their food.

Understandably, this had a profound effect on our ancestors' lifestyles. For the first time in human history, those nomadic hunters and foragers began to settle in homesteads, gradually swapping a diet of meat, fish and mushrooms for one of wheat, barley and peas. While this shift in practice supplied the calories necessary for a steady expansion of the human population, this radically altered the biology of our species for evermore. In many ways, this is something we are still reeling from to this day.[169]

In 1650, a new disease was recognised. The author of the corresponding treatise, Francis Glisson, described 'an irregularity, or disproportion of the parts; namely, the head bigger than ordinary'. The joints were 'less firm and rigid, and more flexible' than usual. Moreover, the 'part of the breast where the ribs meet' was said to be 'somewhat pointed, like the keel of a ship, or the breast of a hen'. This disorder of skeletal growth was rickets.[170]

Sometimes referred to as 'die Englische Krankheit' (the English disease) in German, this was a common disease of industrial Britain, as city smog blackened the skies and poverty restricted the diets of children. Tiny Tim, in Charles Dickens' *A Christmas Carol*, almost certainly had rickets, accounting, at least in part, for his small stature, leg braces and sickliness. Through weakening the immune system, rickets also predisposed sufferers to the most significant infectious disease of the time: tuberculosis. This was a double hit that many failed to overcome. The subsequent prevention of rickets with cod liver oil, a rich source of Vitamin D, confirmed that the most common cause of rickets was, and remains, a childhood deficiency of Vitamin D.[171]

And as I have already alluded to, Vitamin D is the ancient sunshine hormone. Marine and fungal ecosystems are particularly reliant on it, making oily fish and mushrooms excellent sources of dietary Vitamin D. Since the ingested Vitamin D from these foods has already been activated in their non-human hosts, this bypasses the need for activation in the human skin. Come rain or shine, as long as dietary intake remains high, the myriad vital functions dependent on Vitamin D thrive.[172]

Considering all of this, one can finally reason the following. First, our predecessors' hunter-gatherer lifestyle up until 12,000 years ago secured them sufficient pre-activated Vitamin D thanks to plentiful fish, meat and mushrooms in their diets. Second, swathes of ice sheets over higher latitudes peaking about 20,000 years ago restricted our predecessors' advance into the most sun-depleted regions of the Earth. Third, the introduction of farming about 12,000 years ago reduced dietary Vitamin D intake, thereby restoring an ancient reliance on sun-induced activation of Vitamin D in the skin. Fourth, this created a survival advantage for reducing melanin's sunscreen effect, thereby lightening skin tone in some people. Fifth, this survival pressure was greatest at higher latitudes once the melting of ice sheets permitted habitation there.

So, consider this next time you are sitting in the sun. No matter what your skin colour may be, tanning is simply a natural skin adaptation in response to an archaic environmental stimulus: sunlight. This stems from the fact that darker skin blocks the damaging effects of excess ultraviolet rays. On the other hand, the tendency for lighter skin to burn arises from an ancestral selection pressure to maximise Vitamin D activation in areas of low sunshine exposure. And so whatever happens to your skin in the sun, it is, and can only ever be, a reflection of your successful genetic heritage.

I cast a wry smile as I write this, sitting, as I am, in a windowless, air-conditioned hospital office while London bakes in a June heatwave. I smile, because this moment epitomises the barriers our modern civilisation has constructed, both physically and conceptually, in order to shield us all from our natural past. I smile, because the knock-on effects of a rapidly altering environment on the health of its 'wisest' big-brained inhabitants requires much more than hospitals and their windowless, air-conditioned offices. I smile, because there is sometimes nothing else you can do when you reflect on such heart-wrenching irony.

Stage fright

The spotlights singed my skin like unquenchable flames. And yet my palms felt clammy and cold. Four hundred eyes bored holes right through me. Hot under the collar, I swivelled to my right and haphazardly scanned the looming slide on display, grappling for some form of inspiration.

'Errrm...', I repeated several times. I managed a forced, fleeting grin to the audience.

I tugged at my tie-constricted collar. Had someone put the heating on overdrive?

The tiered seats of the lecture hall seemed, one by one, to be loosening their earthly shackles, as though their new calling in life was to tumble down towards the lectern. The camera's red light just in front of me blinked ferociously. I was sure it teased me with a pattern of three short flickers, followed by three longer pulses and then three short ones again. SOS.

Despite reciting 70 seconds of free-flowing speech from memory, the next line melted irretrievably in my mind. Abruptly, my voice fell silent. It suddenly felt as though my carefully crafted mental script might never have existed at all.

The clock ahead of me was now down to 1:30. I'd lost 20 seconds. I could pull this back, I thought. I reached into my jacket pocket to retrieve my emergency notes and hopefully some composure. As casually as I could, I said, 'Well, this is embarrassing. My train of thought has somewhat derailed.'

Some tittered. Others simpered. Those in the audience who had drifted off or been distracted by their phones now clocked on to my unfolding anguish. Most, I hoped, were sympathetic.

I peered down at the four or five bullet points on my sheet of paper, and I tried to identify where my flow had dried up. This unexpected silence was entirely of my own making, and it was claustrophobic.

'During my PhD,' I tentatively went on, 'I p-planned to eavesdrop on the health of n-nerves.' I looked up. The clock was down to under a minute, and I was still on the introduction. It then occurred to me that I had already said that exact line. A drop of sweat snuck its way down the side of my neck and under my collar. I jumped forward to a later bullet point.

'While those initial results are interesting, I'm excited about what comes next...' But I realised, as I said it, that I had skipped the results section altogether ... *and* the methods. The clock stood at 0:35.

I scanned the jumble of words that I *knew* I knew so well, hoping that something familiar would jump out at me; something that might spark a revival. Nothing, though, would return the tens of seconds I had already lost.

'Tracking 25 highly d-devoted p-patients for a year,' I stuttered, 'I showed that ... [0:26, 0:25] ... m-m-muscles ... twitch.' I had lost all threads of cognitive reasoning. My memory had evaporated. For the task at hand, my mind had become nothing but vapour. 0:14.

'In conclusion,' I rushed, 'we are n-navigating this unpredictable p-p-path towards a brighter outlook for p-patients.'

Ending there, I tried to put a brave face on it, but there was no getting around it. I had floundered... big style. Was it the pressure of a larger audience than usual? Was it the blinking red light of the camera? Had I over-rehearsed? Had I expected too much from my fallible brain? Whatever the trigger for such an experience had been, it is the only time I have truly wanted the floor to open up and swallow me whole.

My attempt at progressing in the national Three Minute Thesis competition had failed spectacularly. This contest, founded by the University of Queensland, 'challenges doctoral candidates to present a compelling spoken presentation on their research topic and its sig-nificance in just three minutes'. Up to that moment in April 2019, I had always relished the opportunity to speak publicly on a topic that remains very close to my heart. Luckily, I still do, having subsequently given many other presentations about my research without issue. Paradoxically, this experience has made me a more confident public speaker. The worst had happened, and I had coped. When learning to ski, a fall on the first day relieves the anticipation of fear.

The clamminess and clamming up that I experienced are part of a maladaptive fear response to perceived danger. In this context, 'maladaptive' refers to the errant activation of a biological process that had evolved for a genuine survival-enhancing purpose. My brain was getting me ready for fight or flight. And in order to gauge which option would be best, it caused me to freeze.[173]

My higher order brain functioning became temporarily disabled, including memory, speech and reasoning. My autonomic nervous system took over, pursuing a state of heightened awareness and muscle readiness at the cost of temperature dysregulation and heightened anxiety. Although the threat to my survival as I stood in the lecture hall was purely imaginary, this was enough for the primitive circuits in my brain to trigger its well-rehearsed response to real danger. Abandoning a state of steady composure, my brain shifted itself to a state of tense and paralysing alertness. The status quo had been breached.

Keeping the floodgates closed

This shift from one biological state to another conforms to the central idea in this chapter. Indeed, any biological process is simply a series of constancies punctuated by intervening transitions. While the constancies work hard to reinforce and reset themselves, the transitions step in and disrupt the incumbent stability for something new. Since this new state may well be disadvantageous, unpredictable and/or irreversible, a system that limits the inherent uncertainty of these transitions is likely to perform better overall.

'Homeostasis' stems from the Greek words 'homoios', meaning 'similar to', and 'stasis', meaning 'standing still'. In a biological system, homeostasis is what keeps everything on an even keel. It prioritises balance, dealing harshly and promptly with deviations from its intended state.[174]

Control of core body temperature is one of many physiological examples in mammals. In adult humans, the thermostat in health is set at about 37.0 degrees Celsius, constantly being recorded by a part of the brain called the hypothalamus.[175]

If you enter a cold environment, heat is lost from the body more quickly and the core body temperature drops. Even tiny reductions are exquisitely sensed by the hypothalamus, and a series of well-honed mechanisms are set into motion. Hairs on your skin stand erect to trap heat; muscles activate uncontrollably to generate more heat; blood vessels near the skin surface constrict to divert warm blood to your core. Together, these serve to raise the core temperature back to the target level. In contrast, in a hot environment, vessels near the skin surface dilate and sweat glands activate, both of which are effective at offloading heat. These are self-rectifying feedback loops that operate in both directions to maintain the thermoregulatory equilibrium of health.

Fever, on the other hand, is a biological state that exists in disease. While Hippocrates and Asclepiades, whom I discussed in Chapter 3, considered fever a disease in its own right, today it is more correctly categorised as both a symptom and a sign of a plethora of potential diseases. A symptom is what a patient feels (hot, sweaty, shivery), and a sign is what clinical examination reveals (warm to touch, clammy, flushed).[176]

The shift into a feverish state occurs not because of a failure but rather an awakening of the self-rectifying feedback loops that exist in health. The trigger for this change is an internal thermostat that gets reset to a higher temperature. The same mechanisms that usually operate within narrow boundaries in stable, hospitable environments in health are suddenly cranked into overdrive.

Initially, shivering and the constriction of surface blood vessels are called upon to raise the body temperature to its new target as quickly as possible (>38 degrees Celsius). However, owing to the imprecise nature of this scramble, there is overshoot. This then drives opposing mechanisms, such as sweating and the dilation of surface blood vessels. As a result, the patient experiences alternating phases of feeling cold and hot, as the core body temperature swings back and forth around the raised target.

Undoubtedly, the biological advantage of shifting the body temperature to a less precise state outside its normal range must be worthwhile. This process not only aids host immune function, but also stifles the viral or bacterial invader in its tracks. While ignorant to the mechanics of such an advantage, our distant mammalian forebears who shifted biological states from health to fever (and back again) most seamlessly were more likely to survive the ordeal of an infectious attack and pass on the genes responsible for such success. What this goes to show is that the apparent chaos of fever is an illusion; it is a regulated process serving a clear survival-enhancing purpose.[177]

The state of malignant hyperthermia, however, demonstrates what happens when high body temperatures become unregulated. Most often triggered by the administration of certain anaesthetic drugs during routine surgery, but also by extreme exercise or heat, a dangerous and unstoppable rise in body temperature takes hold. Unlike a normal fever, this is not simply a shift of the usual self-rectifying feedback cycle to a higher centre-point; this is a vicious circle of self-perpetuating feedback loops that know no bounds. Considering this rare disorder is lethal when left untreated, why and how does it happen?[178]

It all begins with calcium. As well as its more familiar role providing sturdiness to bones and teeth, calcium is also a vital component of muscle contraction. When a nerve delivers an electrical signal to a muscle, calcium ions are reversibly released from cellular stores, thereby triggering the mechanics of muscle tightening. This shuttling of calcium ions back and forth is what allows your muscles to do what they do best: to strengthen and relax.[179]

In malignant hyperthermia, the starting pistol is unrestrained release of calcium ions from their stores into the main compartment of the muscle cell. The smoking gun is the presence of mutations in key genes involved in calcium shuttling.

As the floodgates open in this disorder, calcium ions accumulate in muscle cells, leading to uninhibited muscular activation at the expense of the cell's limited energy provisions. Oxygen is depleted, carbon dioxide is generated in abundance and excess heat is produced. The loss of energy reserves leads to the breakdown of cellular pumps and membranes, exacerbating the toxicity. Muscle cells eventually split open, spilling their contents into the blood. The usual blood-filtering mechanisms in the kidneys become overwhelmed and begin to shut down. The heart rate increases in an attempt to deliver more oxygen to floundering muscle cells, while the lungs try to offload excess carbon dioxide. Importantly, the build-up of carbon dioxide acidifies the blood, which applies an extra burden to already failing cells. Eventually, this vicious circle creates such a chaotic state of cellular turmoil that the heart and lungs stop functioning. Hidden among these self-perpetuating feedback loops are theoretical tipping points, which perpetuate the damage until a threshold of irreversible vulnerability is breached.[180]

From the apparent constancy of life, this out-of-control transitionary process, without timely intervention, leads sadly to a new and very different biological state: death.

Next patient, please

So far in this chapter, I have explored how biological systems manage stability and change. From a conceptual look at how early reproductive life may have originated to the bespoke environmental adaptations of lizards and humans over contrasting timescales, we are reminded of Nature's competence time and time again. It does not matter whether you look at bodily temperature control or the delivery of a timed speech to a packed lecture hall, mechanisms exist to keep these biological functions under tight control. Where self-rectifying feedback loops serve to maintain consistency, their rejection in the face of powerful self-perpetuating feedback loops can spell unforeseen disaster.

I paid particular attention to human skin for one principal reason: skin demarcates the inside from the out. Therefore, as the largest organ in the body, it should come as no surprise that it has had a major influence on our adaptation to the environment. After all, every biological system exists within a physical space that cannot be ignored. Therefore, as it is important to do in life from time to time, we have to flip things around and view our observations from another perspective. Imperatively, it is time to consider how the environment adapts to *us*, keeping in mind the same concepts of stability, change and feedback loops. If we don't, we may well end up getting burnt.

In Chapter 3, I discussed how modern doctors typically approach the assessment of a patient. We are taught to listen to the patient's symptoms, examine their bodily signs, carry out informative tests and advise a path of treatment that best fits the most likely diagnosis. Hippocrates taught us this 2,500 years ago.[181]

Now consider the Earth as a patient and apply the same process. The planet may have a different mode of communication, but it remains adequately efficient at getting its message across. Heatwaves, wildfires, droughts, hurricanes and floods are all becoming part of its vernacular, and the frequency of these episodic events is gaining pace. The emergence of a zoonotic virus in 2019 brought modernity to its knees.[182]

On examination, we find rising land and sea temperatures, melting glaciers, thawing permafrost and rising sea levels. Our investigative tools demonstrate fragmented ecosystems, shrinking habitats, wayward soil erosion and imbalanced water and nutrient cycles, all of which are contributing to a species extinction rate that is at least three orders of magnitude greater than in the pre-modern era. This is a biodiversity crisis and a sixth mass extinction in the making.[183]

Considering that these observations clearly point to a transition rather than a constancy, the key question in my mind has always been this: is it anything to worry about? After all, there are many transitions or changes in biology that are deemed healthy. Indeed, change is the premise of natural adaptation, of which there are plenty of examples outlined in this book, and without which we would not observe the outstanding biodiversity of life. Could the planetary changes we are observing reflect the same process?

Or perhaps our assessment of Earth is akin to the mammalian fever, a regulated reset of self-rectifying feedback loops that will revert to its baseline at some point. If that is the case, maybe what we are observing in the world is somehow cleansing, in the same way that a mammal fends off an insidious viral attack during a fever.

Or maybe what our planetary patient is experiencing aligns with the vicious circle of self-perpetuating feedback loops that characterise the fatality of malignant hyperthermia. If so, is our planet on a path towards a threshold of irreversible vulnerability? Has it already passed it?

What is clear to me is that with such a broad range of interpretations and implications, it is vitally important we work out exactly which type of transition we are part of.

But before I delve into this exercise, it is crucial to establish both our perspective and our objectives. If we are simply interested in whether our patient, planet Earth, will survive this transition, then we can answer that right now: it will.

It will undoubtedly end up telling a different story, displaying different signs on examination and giving different results in the tests of the future, but the Earth will still exist in 100, 1000, 10,000, probably even 1,000,000,000 years. Therefore, as a planetary doctor, we would perhaps be better off caring for the other 700,000,000,000,000,000,000 planets in the universe.[184]

However, if instead we care not just about our planet's core survival, but about conserving the *health* of the planet as we know it, then we will have to think a bit harder. First and foremost, I must define what I mean by the 'health of the planet'.

Let me tackle the 'planet' bit first. For the sake of this argument, the planet encompasses the whole of biology and chemistry, as well as the physical spaces in which they operate, including all known and currently unknown parts. It therefore embodies the insects, birds, fish, amphibians, reptiles and mammals, as well as the plants, trees, grasses, fungi and bacteria, all of which make use of the sky, lands, soils, oceans and rivers that Earth has been sustaining, in some form or another, since life began. By specific mention for this exclusively human audience, our own embeddedness within Nature means that we are inextricably reliant on the relative stability of all these moving parts.

And when it comes to defining 'health', I have already offered a rather nebulous description in Chapter 3. I claimed that health, as the absence of dis-ease, reflects a state of ease. At the individual human level, this can be translated into an absence of major or prolonged discomfort. From a doctor's perspective, this generally correlates with a lack of significant symptoms alongside unremarkable examination and investigation findings.

But this is not a particularly useful definition for the purposes of this argument. Instead, we need to know how a state of ease applies to whole populations over extended periods of geological time. My chosen approach returns us to the opening concept of this chapter: successful reproductive life.

Therefore, as I move forward, 'health' denotes the maintenance of mutually sustainable population numbers across the full range of Earth's biodiversity. And conserving that planetary health must be the cornerstone of our onward actions.

An earnest evaluation of Earth

Now that we have established health conservation as a top priority for our planetary doctor, we can explore what our current assessment of Earth tells us.

The first possibility to rule out is that this transition is simply a healthy and natural adaptation, thereby increasing the planet's resilience in the same way the Green Anole lizard adapted to the freezing American winter of 2013/14. The fact that the Earth's sixth mass extinction is already underway is the exact opposite of our working definition of health conservation. Since every other mass extinction has led to a massive change in the make-up of life on Earth, we cannot simply sit idly by, hoping all will continue as before. Because it won't.[185]

This therefore leaves us with two remaining options. Either our planet is in a feverish but regulated state, or it is subject to a series of hyperthermic tipping points that render the situation chaotic at best or irrecoverable at worst. Akin to the conceptual difference between fever and malignant hyperthermia in humans, what this discussion boils down to is determining the planetary balance between self-rectifying and self-perpetuating feedback loops. More simply, is Earth strategically resetting itself to a new norm or is it out of control?

In February 2023, a scientific article was published in the journal *One Earth* that assessed this exact issue. In total, the authors identified 20 physical and 21 biological feedback loops in operation. These loops concern wide-ranging and interdependent aspects of global functioning, involving sea ice, forests, insects, clouds, deserts, peatlands, permafrost and wildfires.[186]

For each of the 41 feedback loops, the effects *of*, and the effects *on*, climate change are summarised side by side in the article. For example, the melting of sea ice caused by higher water temperatures reduces sunlight reflectance (an effect more formally known as albedo). And this means that more heat is absorbed by the darker water that replaces the white ice, which in turn leads to further melting of sea ice.

Worryingly, of the 41 loops identified, two-thirds of them (27) were classified as self-perpetuating. In contrast, only one-sixth (7) of the loops were self-rectifying. For the remaining seven loops, it was not clear one way or the other. One example that falls into the self-rectifying category is how increased rainfall in deserts such as the Sahara leads to the revival of vegetation growth, which in turn boosts the absorption of carbon dioxide, thereby decreasing the overall greenhouse burden.

Of course, feedback loops do not operate in isolation. There is an intricate web of interconnection between these complex global phenomena, which is difficult to fully incorporate within predictive climatic models. It seems instinctive that the more drivers for instability at play, the more unstable the system becomes. But other aspects of these interactions are perhaps not so intuitive.

Models of four real-world systems have been particularly enlightening in this regard. These relate to a fishery on Lake Chilika in India, the collapse of human civilisation on Easter Island, vegetation dieback in the Amazon rainforest and lake phosphorous concentrations. For each model, key drivers that modulate the way these systems operate were assimilated. For example, rates of deforestation by Easter Islanders would have affected the sustainability of food and timber provision, ultimately impacting their long-term survival prospects. Importantly, breakpoints were identified for each model, representing points of irreversible breakdown.[187]

From these four models, three key concepts emerged. First, the detrimental impact of an additional driver was greatest while systems remained furthest away from their respective breakpoints. Second, greater year-on-year variability of the drivers made breakpoints happen earlier. Third, and of most concern, the addition of a second or third driver was often enough to introduce a breakpoint that previously didn't exist, thereby converting a resistant system into a vulnerable one.

Conceptually, this illustrates how the 27 self-perpetuating global feedback loops are likely to interact to destabilise our planet's climatic systems even faster than previously thought.

In January 2023, as director of the Potsdam Institute for Climate Impact Research, Professor Johan Rockström delivered one of the starkest warnings I have heard yet. He emphasised that 9 out of 16 world climatic tipping points 'are showing signs of instability'. Describing the most worrying example of what he calls a 'cascade risk of dominoes', the Greenland Ice Sheet is demonstrating 'accelerated melting, warming four times faster than the planet as a whole; releasing cold, fresh water; slowing down the overturning of heat in the North Atlantic; pushing the whole monsoon system down further south; causing droughts and forest fires over the Amazon rainforest'.[188]

His verdict was simple: our deeply interwoven planetary system is not just unravelling at the seams, but at the hems and cuffs too.

A transition with far-reaching implications

Let's change tack a little and approach this using a concept I introduced in Chapter 2. From what we have seen so far, our planetary transition can be viewed as a series of stressors on a system. At the heart of this, we are interested in whether the health of our planet will ultimately prove its resilience or reveal its vulnerability. The stressful vessel model conceptualises such a problem.

According to this model, individual stressors combine to exert an overall strain on the system. To combat this, alleviating mechanisms attempt to provide stability. In this way, these alleviating measures are similar to the self-rectifying feedback loops that characterise the homeostasis of mammalian body temperature control. Moreover, we learnt how these compensatory mechanisms are fallible when pushed too far, thereby contributing to the failure of any over-stretched system.

The parallels with our planetary transition are hopefully clear to see, but there is one aspect that is perhaps not so obvious. Stressors do not exert their full effect on the system immediately, instead dragging out their impact over variable, sometimes lengthy, periods of time. I've commented before about how this creates a time lag, which can make it difficult to form clear correlations between individual stressors and their resultant strain. Rather ominously, our planetary transition exemplifies this issue perfectly.

Even if all greenhouse gas emissions stopped overnight, one climatic model published in 2020 showed that after a welcoming initial drop in global temperatures, a counter-intuitive and self-perpetuating rise was predicted to dominate from the year 2150. The main reason is that the melting of sea ice, through its unstoppable knock-on effects on sunlight reflectance and permafrost thawing, takes time to exert its full impact. This lag, according to the model, would convert an initially favourable outcome into an unavoidably catastrophic one.[189]

I acknowledge that the evidence I have put forward so far paints a pessimistic picture. It suggests that the balance of global feedback loops is shifted firmly in favour of the self-perpetuating, unfavourable kind. But does this mean the health of our planet is doomed? Not necessarily. However, it *is* cause for considerable concern, and it *does* require us to somehow shift the balance back towards a dominance of self-rectifying feedback loops. One way to do this is through reducing the stressors we place on the planetary system, permitting any exhausted alleviating factors to kick back into gear. In that way, our planet may instead go through a phase of 'chronic fever', which I would certainly consider the lesser of two evils.

I conclude this chapter with an explanation of its enigmatic title, The Climateric. This is a neologism that is closely related to the word 'climacteric' (note the omission of the middle 'c'). Climacteric can be defined in two main ways. As a noun it means a 'critical period or event', and as an adjective it means 'having extreme and far-reaching implications or results'. I therefore propose that the climateric, as a slight tweak on this word, is the perfect representation of the climatic transition we are now observing on Earth. After all, isn't the rapidly changing health of our planet a sufficiently critical event with the most far-reaching implications that, at the very least, it deserves its own word? Who knows, maybe an appropriate level of respect for what's happening will then follow.

CHAPTER 5

Our Bipolar World

Dualism is conflict's most doting custodian

There is an unspoken rule among frontline hospital personnel. The first time I witnessed its real impact was when I worked in one of the busiest acute hospitals in England. During the first 12 weeks of 2013, there were 67,109 Emergency Department attendances at the now disbanded South London Healthcare NHS Trust. This was the seventh highest total among 186 English NHS Trusts, equal to around 5,500 attendances per week across two acute hospitals. This was just shy of 800 patients per day.[190]

One of the acute hospitals in the Trust was Queen Elizabeth Hospital in Woolwich, which was where I spent my third year as a doctor in 2012–13. For four months during that winter, I worked in the hospital's acute medical unit, which is a 78-bedded unit admitting those patients from the Emergency Department in need of hospital-based treatment. This type of unit had been set up specifically to look after patients with a range of heart, lung, bowel and brain problems. Treatments ranged from oxygen and nebulisers for the lungs to fluids and medications infused directly into veins.

The unit was a revolving door. Within 72 hours, patients typically either returned home after satisfactory treatment or were transferred to one of the long-stay wards for more specialist care. Given the high volume of patients passing through the unit, this was a time when my independent clinical decision-making matured in earnest.

There were, nevertheless, natural lulls in the flow of patients being admitted. It was at these times when the unspoken rule became a law unto itself. All it took was one careless mention of the 'Q-word', perhaps some casual remark such as 'The on-call seems *quiet* today.'

The rule was simple: even if the on-call was quiet, and even if everyone was thinking how quiet the on-call was, *under no circumstance should anyone vocalise it.*

If you happened to break this rule, as I irrepressibly did once or twice, you risked feeling guilty about the inevitable resurgence in hospital admissions that followed. The word 'quiet' was treated as the quasi-evil work of some health demon, capable of summoning illness out of nothing. So how had the mention of a seemingly innocuous word become so taboo in frontline healthcare circles?

This observation, of course, has a rational explanation. It is based on the fact that extreme occurrences in a variable system naturally regress towards the average. For example, when the on-call take had admitted zero patients over the previous two hours, the *only* possible change was for more patients to be admitted. So when someone said the word 'quiet', it was simply a proxy marker of that unusual circumstance. The result was inevitable, not because of any demonic influence, but because it was the only direction in which the system could move.

In some cases, this effect facilitated rapid swings from one extreme to the other. One hour there may have been insufficient work to occupy one person, but during the next hour even three people struggled to get everything done. This is an effect that continues to plague on-call rota coordinators within any modern healthcare system. How do you design a system that is flexible and cost-effective enough to deal with the unpredictability of acute illness? After all, most people do not plan to become unwell.

Unfortunately, in stark contrast to the let-ups and the superstition that surrounds them, there are occasions when pressures on acute healthcare systems mount to breaking point. We saw this during the COVID-19 pandemic, particularly in countries running out of particular resources. I remember seeing news reports at the time about insufficient oxygen canisters in Brazil and overloaded critical care units in Italy and Japan. In many countries, including the UK, the scarcity of personal protective equipment and blood tubes severely hampered the safety and quality of the care that could be provided.

Even before the COVID-19 pandemic, though, hospitals and their staff had been forced to manage a variety of major incidents. In 2017 alone, five acts of extremism on UK soil resulted in a total of 41 deaths, leaving hundreds of others with physical injuries and psychological trauma. Even more devastating has been the incomprehensibly huge number of deaths attributable to violent natural processes, such as the Indian Ocean earthquake and tsunami in December 2004, which killed 227,898 people in 14 countries. Horrifyingly, extreme events like these bulldoze almost all efforts of preparatory mitigation.[191]

We live in a world where extremes will always occur. From a pure statistical standpoint, these rarer events happen owing to so-called noise around the average. Since a completely noiseless system is impossible, we are all accustomed to a degree of variation within manageable limits. Variety, after all, is the spice of life.

But it is also true that, beyond a certain point, the noisier a system behaves, the more hostile it tends to become. While extremes at either end of the spectrum are to be expected from time to time in any naturally variable system, including acute healthcare, we generally fare better when faced with *fewer* and *less extreme* extremes. In this chapter, I explore how this concept can be applied to our health, our planet and the natural world.

A disease of extremes

Since beginning my training as a neurologist in 2014, my on-call commitments have focused on patients suffering from brain and nerve problems. Among the thousands of patients I have observed, there is one diagnosis that epitomises the extreme manifestations of disease. Amazingly, first reports of this disease only appeared fairly recently in 2007. Since then, it has rapidly become the most well-understood condition among a broader family of related disorders.[192]

This knowledge has dramatically altered the way we now approach certain patients presenting with overlapping neurological and psychiatric symptoms. It is the perfect example of how variable, seemingly unconnected accounts of an undiscovered disease have been condensed into a precise unifying narrative, like water vapour forming the morning dew.

This disease mainly affects young adults, occurring four times more commonly in women than men. Its course can be divided into five key phases, as it yo-yoes between extremes. First, there is a non-specific phase of headache, fever and fatigue lasting days to weeks. This then abruptly gives way to a psychotic phase, characterised by hallucinations, delusional thinking, memory lapses, compulsive tendencies and mood disturbances. Most also develop epileptic fits. However, within two weeks, these predominantly psychiatric symptoms are usurped by an unresponsive phase, the hallmark of which is something called catatonia. Rather unlike the 1990s Welsh rock band of the same name, this term describes a triad of no speech, no movements and no interaction, all while eyes remain open and vacant. Gradually, this unresponsive phase switches into the penultimate phase of the disease. This is characterised by excessive movements of the mouth, tongue and fingers, in conjunction with potentially lethal swings in heart rate, blood pressure and body temperature. Finally, a recovery phase usually predominates, leaving around 80% of treated patients with no or only mild sequelae. Sadly, even with treatment, 20% of patients either die or suffer long-term disability.[193]

With such a rollercoaster of serious disease manifestations, I am sure you are wondering what on earth could cause it all. The answer rests on a betrayal, albeit an accidental one.

The immune system exists to defend its host from attack. In order to achieve this noble aim, it has a variety of tricks up its sleeve. Its army of white blood cells engulf, puncture and dissolve their bacterial, viral and fungal victims, while the same cells also eliminate cancerous cells on the inside and polluting particles from the outside. Like the Asian elm tree subduing the fungus that causes Dutch elm disease, as discussed in Chapter 2, the human immune system generates baseline and honed responses. These are the innate and adaptive divisions of the immune system, respectively.[194]

Most relevant to our discussion here is the adaptive response. As a critical part of this process, specialised immune proteins, known as antibodies, are produced. Each type of antibody is constructed so that it latches onto a specific target, called an antigen. If you ever took a rapid home-based COVID-19 test, then this was based on the detection of the virus' specific antigens in your nose and throat. Think of an antigen as a protein's unique fingerprint and the corresponding antibody as a precise map of its most distinguishing features.[195]

There is no doubt that antibodies form an essential line of defence against invading bacteria and viruses. But sometimes the immune system gets it wrong. While its capacity to produce antibodies against invading proteins works well in averting infectious diseases, this same capacity has the potential to unleash an armada of self-destruction.

A brain under attack

We call this type of disorder an autoimmune disease. And when neurological and psychiatric manifestations predominate together, as they do in the description of symptoms I've just given, we more specifically call it an autoimmune encephalitis. In this family of conditions, the immune system produces antibodies targeting particular antigens found on the surface of neurons in the brain. Thankfully, these conditions are rare.[196]

The immune target that leads to the yo-yoing phases of symptomatic extremes I've described is a cellular receptor that is pivotal to communication between neurons. This receptor is controlled by the main excitatory chemical in the brain, known as glutamate. When activated by glutamate, this receptor opens up a channel for calcium ions to flow into neurons, thereby governing a series of vital functions within the cell. When antibodies attack this widespread brain receptor, it is little surprise that it causes such chaotic brain outputs.[197]

So, breaking down the term 'autoimmune encephalitis', we can begin to understand how it got its name. 'Auto' means 'self' – think *auto*matic, *auto*pilot and *auto*biography – and 'immune' refers to our body's system of 'attack'. 'Encephal-' happens to refer to 'brain', while anything ending in '-itis' – think appendic*itis*, conjunctiv*itis* and tonsill*itis* – indicates which part of the body is affected by the disease.

Therefore, autoimmune encephalitis is just a fancy way of saying 'self-attacking brain disease'. But if doctors called it that, then everyone else would know exactly what they meant. And we couldn't have that, could we? Plus, quite frankly, it sounds way too scary when expressed in those everyday terms.

Earlier, I claimed that this disease is caused by an accidental betrayal. While the 'betrayal' part should now be obvious, given the brain is attacked by the very system that should be protecting it, the 'accidental' descriptor is less clear. Why, then, does the immune system suddenly develop such destructive auto-antibodies targeting the brain in these cases?

Part of the answer relies on the brain exhibiting something known as immune privilege. This phenomenon was first appreciated in a series of animal experiments in the 1920s and 1940s. These studies showed that while donor cells grafted under the host's skin were quickly rejected by the host's immune system, those placed in the animal's brain were simply ignored. Somehow, the brain was not subject to the same immune surveillance as the rest of the body. Intriguingly, though, when cells were transplanted into the brain *after* other cells from the same donor had been put under the skin, they were no longer ignored, prompting their subsequent rejection from the brain.[198]

This was a curious set of results. On one level, the brain appeared to be invisible to the immune system, capable of concealing stowaways from any patrolling units. Yet after discovering external evidence of this oversight, immune cells were proficient at disbanding the unwelcome visitors from the brain. So what exactly was going on? And what relevance does this have to our current understanding of autoimmune encephalitis, and perhaps more widely?[199]

When it comes to your bodily organs, the brain has a unique set of attributes. For one, it is housed in a durable, bony casing. But while the skull serves as sturdy protection from outside trauma and potential intruders, its inelasticity copes poorly with pressure rises on the inside. Were the immune system to suddenly summon a flood of circulating cells and blood within the skull, many of the resident brain cells living within those unforgiving walls would simply be squeezed to death. For this reason, strong survival pressures during evolution would have favoured a less vigorous immune response in the brain. But this would only work if the brain could be protected from invasion in other ways.[200]

Thankfully, integral to our brain development was the specialist blood–brain barrier, which I introduced briefly in Chapter 1. While the primary function of this barrier is to ensure an optimal chemical environment for normal brain function, it also resists potential invasion from the bloodstream. In combination with the impervious skull on the exterior, this interior barrier sanctions the brain to do away with the body's usual mechanism for immune patrol.[201]

Further proof that the brain evolved a very different relationship with the immune system than other body organs comes from its lack of a draining lymphatic network. I am sure we have all felt our own necks for swollen lymph nodes while harbouring a sore throat. While this is evidence that our immune system is busy recognising and targeting unfamiliar antigens collected from our throats, this doesn't happen in the brain. The brain's lack of lymphatics means antigens simply escape the same level of scrutiny.

The big reveal

All this talk of privilege, concealment and betrayal sets us up nicely for this story's denouement. Earlier, I said that this type of autoimmune encephalitis occurs four times more commonly in women than in men. The main reason for this disparity between the sexes can now be revealed.

It turns out that just over half of the patients with this disorder also have an ovarian tumour. As the mammalian egg-producing organ, the ovary contains generalist cells that can transform into any specialised cell in the body, including those found in bone, hair, teeth and, of most relevance here, the brain.[202]

Therefore, when an ovarian tumour grows, it can contain a jumbled mix of all these various cell types. But unlike the brain, the ovary does not exhibit immune privilege. And so this type of tumour presents an array of novel antigens to the immune system. Crucially, this can include the widespread glutamate receptor so critical for neuron-to-neuron communication in the brain.

The staggering conclusion is that this natural occurrence has the same effect as those early experiments in the 1940s in animals. The immune system is sensitised to the targeted receptor by its chance appearance in the ovary, acting as the perfect trigger for an extremist attack on the brain.

The accidental betrayal is complete.

It follows, rather surprisingly, that one of the most useful tests neurologists can recommend in patients with this condition is nothing to do with the brain and nerves at all. Instead, we prioritise ultrasound imaging of the ovaries. When a tumour is found, it must first be removed by our surgical colleagues for any hope of neurological recovery.

When I first learnt about this rare disease, I was in respectful awe. The fluctuating extremes borne out in affected patients were so all-encompassing yet seemingly illogical. I already knew the workings of the human body were fascinating, but what struck me more than ever was how inherently connected its various functional parts are. It has undoubtedly made me ponder on the wider implications of this. If irregular cell growth in the ovary can stir up such turmoil in the brain, thereby linking two parts of the whole that at first glance appear disconnected, then which other connections are we currently oblivious to?

One of land's many ends

Kanyakumari lies at the southernmost tip of mainland India. Uniquely situated at the confluence of the Bay of Bengal, the Arabian Sea and the Indian Ocean, it doubles as a site of great geographical interest and a place of holy pilgrimage for Hindus. This was where Kanya Kumari, the virgin incarnation of the Supreme Goddess, is said to have overthrown the powerful demon king Banasura.[203]

Legend has it that Banasura had wrought tremendous suffering in the region through his greed and domination. Of ominous note, his actions had dispersed the gods, who usually oversaw the natural components of life. Unsurprisingly, this had led to a state of natural disorder for the citizens of the time.

Owing to a special privilege bestowed upon Banasura, only a virgin goddess could conquer him and restore the rightful order of the gods. However, the only suitable candidate for this noble task, Kanya Kumari, was due to marry Lord Shiva, the Supreme Being. Recognising that this union would relinquish the young goddess' virginity and make her ineligible to perform her ordained duty, the gods tried to break off the wedding.[204]

They bade the groom-to-be to produce three exceptional items as the price for his bride. These were 'a sugarcane stem without rings, a betel leaf without veins and a coconut without eyes', all unnatural items concocted by the gods to ensure Shiva would falter. But their plot was to no avail, since the powerful Shiva was able to create all three of the items without difficulty. And much to the gods' chagrin, the wedding was set for dawn on the chosen day.

But the gods conjured up a final trick, this time proving successful. They caused a cockerel to crow earlier than usual on the day of the wedding, confusing Shiva and leading him to believe he had missed the dawn ceremony. As a result, he never even turned up to his own wedding.

Jilted in such a manner, Kanya Kumari became understandably angry, turning the wedding banquet's rice into the sand 'of many different colours and shapes' that blankets the town's beach today. Despite this, the gods were overjoyed when Kanya Kumari finally deployed her wrath to kill Banasura, thereby removing the menace that had wreaked havoc for so long.

Happily, the anticipated restoration of natural order soon followed, and the town's 3,000-year-old Kumari Amman Temple still commemorates this triumph of good over evil today.

I was fortunate enough to visit Kanyakumari with three of my friends from medical school in July 2005. We were on a month-long placement after our first year of university, attempting to gain insight into world medical practices.

Although a much younger version of myself failed to fully appreciate the spiritual significance of this site at the time, it was not difficult to marvel at its geographical wonder. Thinking back to the tricolour appearance of the converging seas and ocean, as that blue vastness encircled the lonely tip of land on which I stood, I fondly reflect on that journey, taken so many years ago now.

There is no doubt that stops in Antarctica, Mozambique and Madagascar, before sailing across the ocean to that amazing location, had made the world a far more connected place than I'd previously imagined. And the completion of this journey in what seemed like a relatively short space of time only amplified its impression on me. Quite simply, it was a trip that could never be repeated. It had been, dare I say it, not just the trip of a lifetime, but one worthy of many lifetimes.

But before I go on, I must come clean about something. For I have been deliberately misleading you. This journey I am describing is not one I have ever taken. Indeed, it is not a journey *any* human has ever taken. I can be so sure because the exact journey I have in mind began way before our species ever came into being.

A journey of extremes

Approximately 170 million years ago, the configuration of the world's landmasses looked very different. Instead of the seven continents we instantly recognise today, there were two supercontinents known as Laurasia in the north and Gondwana in the south. But this was about to change, as Gondwana was starting to fracture.[205]

The forces driving this fragmentation have been continually reshaping Earth's landscapes and waterscapes for at least 3.4 billion years. As geological events tend to unfold extremely slowly, we are easily deceived when we observe the seemingly static nature of our continents and oceans today. And the scientific theory that has honed our understanding of this fascinating process is that of plate tectonics.

Tectonic plates make up the Earth's crust. Essentially, they are thin strips of cold, solid rock at the Earth's surface, sitting on a much thicker, much hotter and more dynamic mass of rock known as the mantle. Each plate is like a wooden raft on top of a sluggish sea of marbles, permitting its movement in whichever direction the prevailing forces deem fit.[206]

And where two plates come into contact, three types of interaction can occur. If the two plates separate, then molten rock from the mantle rises through the gap, solidifying as new crust and forming a rift valley on land or a mid-oceanic ridge under water. If the two plates collide, then as one plate dives beneath to be churned up by the hot mantle, the other climbs atop to form a vast mountain range. Finally, the two plates may do neither of these things and instead grind alongside each other, producing intermittent earthquakes.

By about 70 million years ago, as Gondwana was gradually splintering into its constituent parts, one portion appeared to get a particular spurt on. For the next 20 million years or so, this hasty tectonic plate seemed to speed northwards across its mantle bed at about 18 cm per year, far exceeding the usual tectonic pace of under 10 cm per year.[207]

It turns out that this anomalous behaviour was that of the Indian plate hurtling towards its current location. It is remarkable to think that for hundreds of millions of years the southern tip of India was nestled in the middle of Gondwana, near to land that would eventually assume the current positions of Antarctica, Mozambique and Madagascar. It is perhaps even more remarkable that the whole of India (with Sri Lanka and the Indian Ocean in tow) travelled across the then-present Tethys Ocean before colliding with the Eurasian plate, itself formed when Laurasia split into North America and Eurasia.[208]

Making impact around 50 million years ago, the collision between the Indian and Eurasian plates was so powerful that it produced the Himalayan mountain range, whose peaks continue to rise by 1 cm each year. This is driven by the persistent movement of the Indian plate by 5 cm per year, progressing northwards at almost the same rate as the average child grows taller.[209]

So the journey I reflected on, as I recalled my 2005 trip to Kanyakumari, was one taken not by me or my friends, but by the land itself. Spanning 170 million years and 6,000 km, it was without doubt a journey of extremes. It had contributed to the break-up of a super-continent that had at one time covered a fifth of the Earth's surface, while also leading to the formation of the tallest mountain range on Earth. It also gave rise to the fastest tectonic plate speed ever recorded.

Or did it?

Revising your assumptions

Until very recently, interpretation of the available data confirmed the superlative pace of the Indian plate as it travelled eastwards and northwards. By examining the ocean floor and looking for tell-tale reversals in the Earth's magnetic field over millions of years, scientists have been so confident of this finding that the only remaining question was 'why?' According to the leading theory, the Indian plate benefited from a forceful volcanic push from behind.[210]

However, in an unexpected twist, new analysis based on high-resolution data has revealed a competing interpretation, one that has challenged the very existence of this extreme observation in the first place.[211]

To understand how this surprising conclusion came about, I must first highlight two important facts. First, the volcanic activity thought to have propelled the Indian plate along its route was located near the border between the African and Indian plates. The African plate was on the western side of this volcanic plume and the Indian plate was on the eastern side. Second, on the opposite, western edge of the African plate, the mid-Atlantic oceanic ridge was slowly forming owing to the separation of the South American and African plates. Crucially, therefore, we know that both the African and Indian plates were moving in the same direction: eastwards (Figure 5).

For the volcanic push theory to hold true, its push should have radiated out equally in all directions. Therefore, it should not only have *accelerated* the dispersing Indian plate as it moved eastwards, but it should also have *decelerated* the approaching African plate.

But that's not what the new analysis found. Instead, it showed that *all* the plates in the vicinity accelerated in whichever direction they were already moving. Just as the volcanic activity was ramping up about 67 million years ago, why was it that the African plate was being *attracted* to it, while the Indian plate was being *repelled* by it? This just didn't make sense.[212]

In science, there are times when a new theory is needed to explain the data in front of you. In rare circumstances, a complete re-evaluation of the problem is required. Faced with a clear conflict between their new analysis and the leading theory, this is exactly what the authors of this new study were forced to do. Their brand-new theory was to blame an honest mistake.[213]

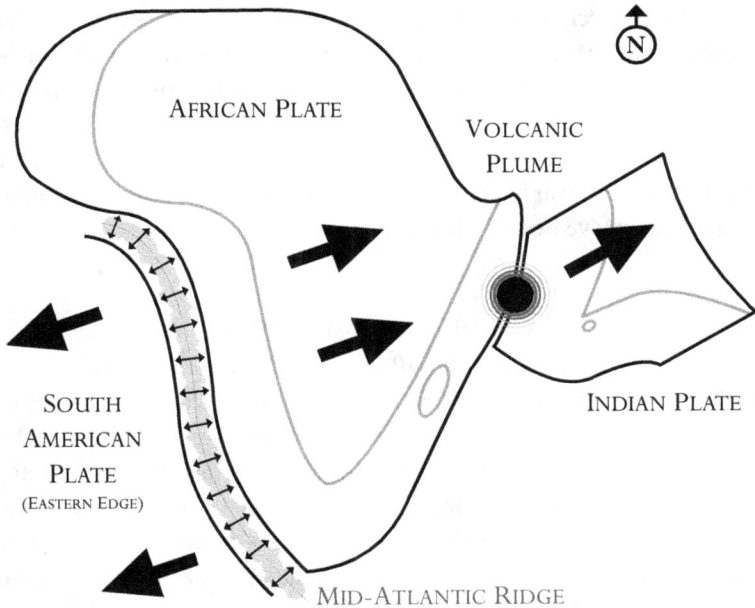

Figure 5. Volcanic plume theory: depicting the dynamic relationships between the Indian, African and South American tectonic plates about 67 million years ago.

We can probably all recall that the speed of an object is equal to distance over time. But let's assume for a moment that the extremely high speed of the Indian plate had been calculated wrongly. This would mean that either the distance measured was too high or the time recorded was too low (or a mixture of the two).

Given that every single plate in the new analysis was shown to accelerate by roughly the same percentage, then it is likely that the same mistake had been applied across the board. When dealing with events that happened so long ago, any estimation of time is fraught with challenges. This makes it much more likely that the potential error was due to an underestimation of the time, rather than an over-estimation of the distance.

The authors applied this logic to the numbers in front of them. All the plates appeared to accelerate during a critical period of time lasting 2.9 million years. But what if the original estimation of that period's duration was wrong?

When the scientists adjusted the duration of that period so that it lasted two-thirds longer than originally estimated, all the plate speeds returned to their long-term averages. This now made much more sense. The correction of this mistake had resolved the conflict. Ultimately, the maximum speed of the Indian plate was so irregular that it probably never happened at all. This was the scientific equivalent of realising you've been barking up the wrong tree.[214]

The moral of this story

So why have I gone to such lengths to share this honest mistake with you? After all, the façade of knowledge often benefits from a spot of remodelling, doesn't it? How could this particular mistake ever be relevant to the interdependence between human health and Nature?

I will first tell you what my motivation is not. It is not to make any scientist feel silly, smug, small, big, sorry or proud. Science, in its purest sense, represents the cold, hard truth. Science is not a competition. Science has no ego. Science has no emotion. Any recognised errancy along the way only adds texture to its unerring wonder. Only science knows what is around the corner. Science doesn't care who is right or who believes it. Science is not bothered whether it makes you happy, sad, relieved or angry.

But there is a deeper, more heartfelt message here too. A scientist's quest is to question *everything*. This will inevitably lead to scenarios, such as the one outlined here, where the top theory concerning the maximum speed of the Indian plate is challenged. Given such a strong counter-argument materialises at all, based on an ostensibly simple mathematical error, you may instinctively feel that this undermines the whole scientific method.

For me, it does the exact opposite. It reinforces the ingenuity of the scientific method to keep asking the same questions in different ways. By extension, it cements the untold number of examples where the consensus only strengthens as time goes on. If scientists weren't constantly on the lookout for mistakes, or if they simply overlooked conflicting results as curious quirks, then how could we ever trust those interpretations that are reinforced time and time again?

This then brings me onto my final point of this section. The study of plate movements millions of years ago may initially appear to be an abstract pursuit with little pragmatic value today. Arguably, however, it has a far greater connection with Nature and the health of its constituent parts than anything else.

Without plate tectonics, life as we know it would almost certainly not have evolved. Heavily influenced by the geological drama of volcanic eruptions and asteroid strikes over billions of years, the restless movements of the plates beneath our feet have provided the environmental variety, with its inescapable extremes, for life to not only initiate, but at times to either thrive, flounder or recuperate.

The specific journey of a landmass traveling across an ancient ocean around 67 million years ago may seem inconsequential to our plight today, but I would contest that strongly. For one, the Indian plate itself hosts around 1.5 billion people today. Moreover, a quarter of the world's population (equal to 2 billion people) rely on life-sustaining meltwater from the Himalayan mountain range, produced by the collision between Indian and Eurasian plates 50 million years ago. Even as recent as 2004, the Indian Ocean earthquake and tsunami that killed almost 230,000 people and immeasurable numbers from other species was due to the lurching movement of the Indian plate as it was forced beneath the neighbouring Burma plate.

These consequences of plate tectonics, both good and bad, are part of the wider workings of our planet; a set of physical principles that continue to set the make-up of our lands, our oceans and our atmosphere in the twenty-first century.

For example, the surface cooling of volcanic lava exposes fresh rock to the atmosphere, which reacts with and removes carbon dioxide from the air, thereby cooling the planet. At the same time, the spewing of volcanic magma fills the atmosphere with carbon dioxide and sulphur dioxide, crafting a delicate balance between the warming effects of carbon dioxide through greenhouse heat-trapping and the cooling effects of sulphur dioxide through sunlight reflection back into space. One cannot escape the fact that the Earth's climate is constantly shaped and disturbed by these processes (and many others), usually over gargantuan timescales.[215]

In the excellent BBC documentary by naturalist Chris Packham, entitled *Earth*, there is a clear focus on the extreme episodes in our Earth's long history. In a written interview about the series, Chris Packham summarises this perfectly: 'From being devoid of atmosphere to having a blue sky; from being warm at the Poles to being frozen to the equator; from virtually being covered in trees to being completely barren; from life being only in the sea to life smothering the land – we've looked at those extremes.'[216]

When we think back to the Hindu legend of Kanya Kumari, it is clear that Earth's geology has profound effects on our modern worldly perceptions. But despite this, there is one thing we should not forget. While it is the extremes that are often of the greatest appeal to our imaginations, our Earth is a beautiful, awe-inspiring and relatively stable mass of rock, water and gases, deserving all the care and respect we can muster. I, for one, hope we can keep it that way.

A year of anomalies

This chapter pays respect to the ups and downs of life. Whether we reflect on the quietude and chaos, the good and evil, or the mountains and deep-ocean ridges, we are often left with the startling realisation that these contrasting phenomena are inextricably connected. Although the basis for these connections may emanate from superstition, spirituality or science, each is a story with one simple purpose. It exists to be told.

When we consider the story that planet Earth is telling us, I introduced the climateric in Chapter 4 to describe the critical period of climatic extremes we are currently observing. I framed the problem in the context of transitions and constancies, which is analogous to this chapter's focus on extremes and how they bookend any interluding stability. But there is one pair of extremes that deserves special mention: the Earth's North and South Poles.

So what exactly is the story of our bipolar world?

Climate Reanalyzer is a good place to start, as it helps to set the scene. This is an openly available online resource produced by the Climate Change Institute at the University of Maine, USA. Based on data from standardised thermometers and satellite images across the world, it provides daily updates on air and sea temperatures, as well as Arctic and Antarctic Sea ice extent, all portrayed using interactive time series and map animations. On 29 August 2023, I took a look.[217]

The first parameter I clicked on, the 'Northern Hemisphere Sea Ice Extent', showed the natural rhythm of the seasonal extremes. On the day I looked, sea ice extent was on its usual decline towards its expected annual low in mid-September. After all, we were approaching the end of the Northern Hemisphere summer.

But as I looked more closely, and hovered over each year going back to 1978, a worrying trend became apparent. From about 2007 onwards, both the sea ice maximum in mid-March and the minimum in mid-September were falling. The Arctic was not only melting that bit more during the summer, but it was also failing to re-freeze to the same degree each winter.

I then clicked on the button that said 'switch hemisphere'. First, I had to recalibrate. Although the natural swing between seasonal extremes was still there, this graph was a mirror image of the first one. The maximum sea ice extent in the Antarctic is instead observed in mid-September at the end of the Southern Hemisphere winter, while its minimum is in mid-March after its summer. As expected, the two hemispheres were opposites of each other.

But there was something else on this graph that defied expectation. Oddly, the 2023 line was drifting away from the main pack of squiggles like a dislodged icefloe. The Antarctic Sea ice was still reforming during the Southern Hemisphere winter, but nowhere near as quickly as in each of the previous 45 years. This looked like an anomaly.

When I moved on to the other graphs displayed on the website, this perturbation was consistent with anomalously high sea surface and air temperatures. To put it succinctly, 2023 was witnessing the most extreme values since records began in 1978. The polar ice caps were feeling the heat.

Hang on, though, doesn't this seem familiar? Before we go barking up the wrong tree, couldn't this be another example of our speeding Indian plate 67 million years ago? Couldn't it all be a mistake?

The first thing to say is that it is much easier to record something happening *now* than something that happened millions of years ago. It is almost impossible that all the thermometers and satellite images across the world are getting it wrong in parallel.

That aside, a more conclusive way to approach this possibility is by seeing if our basic parameters of the Earth's climate, including 2023's anomalous sea ice extent, sea temperatures and air temperatures, corroborate with more complex aspects of meteorological function. More fundamentally, while these average measures of our climate may be irregular, does this alter the weather patterns we experience from day to day? To help answer this, we must temporarily ignore Adam McKay's 2021 directive and look up.

The power of jet streams

Achieving speeds of up to 275 miles per hour and travelling at altitudes at least as great as the tallest mountains on Earth, four jet streams propel air from west to east, thereby governing hemispheric weather patterns. With each hemisphere boasting a polar and a subtropical jet stream, we owe more than we might think to these floating conveyor belts of wind.[218]

Their unidirectionality stems from Earth's constant eastward rotation, pulling its blanketing atmosphere of gases with it as it spins about its polar axis. The force driving the relentless flow of the jet streams is simply a matter of energy dynamics. As energy builds up at the equator owing to the sun's intense heat, it travels down temperature gradients to the cold poles. When the difference between the equatorial and polar temperatures is at its greatest, which occurs in each hemisphere's winter, the jet streams not only speed up, but they also straighten up. Unsurprisingly, the converse happens in the summer. And as the jet streams slow, they begin to meander off course like someone who has had one too many drinks after a summer festival. It follows that a slower jet stream is a wavier jet stream.[219]

While these seasonal variations are well-characterised and undisputed, the same effect is now being observed owing to those anomalous temperatures I discovered on the Climate Reanalyzer website. As the poles are disproportionately warming up, fuelled by the self-perpetuating loops I discussed in Chapter 4, the gaps between equatorial and polar temperatures are shrinking. This exacerbates those slower, wavier jet streams that characterise the summer months in each hemisphere. And what this does is it increases the chance of a jam.

We've all been there, haven't we? You're making good progress on the motorway, and suddenly you're confronted by a wall of red lights ahead. You brake to a near stop and fear a lengthy wait because of unforeseen roadworks or a recent collision. But no, within a few minutes, the flow of traffic picks up again without any evidence of an obstruction. You're left relieved, but also a little bemused. In fact, you were just part of a phantom traffic jam, or jamiton.

This is a well-understood phenomenon that happens when the density of cars on the motorway exceeds a certain capacity. Above this threshold, the fast-moving cars are sufficiently close together that the smallest reduction in speed by one car has an amplifying effect on all the cars behind. This expanding ripple quickly leads to a temporary slowdown, or even a standstill, allowing the density of cars ahead to return below the critical capacity threshold.

It has been suggested that the same phenomenon explains the tendency for slower, wavier jet streams to jam. As they meander more and more towards the equator and poles, the capacity for constant flow is breached, and a jet stream jamiton ensues. The issue is that this blocking pattern allows extreme weather patterns to build up, lasting weeks in many cases. This is what caused the development of heat domes over North America and southern Europe in the summer of 2023, contributing to both the hottest month globally and the largest European wildfire ever recorded.[220]

So far, then, it seems that those climatic anomalies in sea temperature, air temperature and polar sea ice extent are indeed associated with extreme weather events across the globe. But as I have only really dealt with the issue from the air, it's time to swap this aerial perspective for a marine one.

The pumping hearts of Earth's oceans

In some places, such as Kanyakumari at the southern tip of India, you can begin to appreciate the interconnectedness of Earth's bodies of water. But while the blending of teal, azure and cobalt is plain to see on the surface, what you don't see is what goes on in the depths.

All of Earth's oceans are deeply connected by large, slow-moving underwater currents that help to shift heat, salt and nutrients around the globe. Marine and terrestrial life are reliant on them. The weather patterns we experience from day to day are influenced by them. And yet, before researching this chapter, I admit I knew very little about them.

The main drivers for these continuous undercurrents are twofold, and again they depend on gradients between extremes. The first one draws parallels with the formation of the aerial jet streams. In the same way that the sun warms the equatorial air, it also heats the water, thereby setting up a temperature gradient between the equatorial and polar regions of the oceans. In the Atlantic, this drives the Gulf Stream, as water moves northwards in between Central America and the Caribbean islands, before flowing across the Atlantic Ocean to North-West Europe. As it travels further northwards, it cools sufficiently to contribute to the formation of the Greenland Ice Sheet.

Consequently, it establishes the second type of gradient that drives these vast underwater rivers. This time it is a gradient based on salt.[221]

As sea water freezes to form the Greenland Ice Sheet, it leaves behind its salt, making the fluid water that remains even saltier. Since saltier water is heavier, it slowly sinks several kilometres to the deep ocean. This is then drawn southwards back towards the equator and beyond to make space for the ongoing descent of dense, salty water. Combined with a similar process near the Antarctic in the Southern Hemisphere, the resultant deep-flowing currents circuit the globe, eventually ascending back towards the surface in the Pacific and Indian Oceans.

This thermohaline circulation, as it is known, effectively has two types of pump. It is as though the oceans have two slowly beating hearts, keeping the planet alive. The first is a horizontal, thermal pump that shifts the warmest surface water away from the equator, while the second is a vertical, saline one that sinks the densest water to the depths.

Despite the elegance and grandeur of such a system, scientists have recently noticed a potential hitch. In the same way a weakened pulse might disclose a failing heart, scientists have picked up on a tell-tale sign that this almighty circulation might be slowing down.[222]

I say 'might', because it is a difficult thing to be sure about. It could be an early warning sign of something more serious, or it could reflect natural noise in the system. Working out this distinction has quickly become the hottest topic in the field.[223]

Except it actually concerns a cold spot. More specifically, it concerns an area of cooler water around the Greenland Ice Sheet in the Northern Arctic region.

But wait a second, didn't the Climate Reanalyzer website show us that sea surface temperatures are on the rise? While that, as an average across all the oceans and seas on the planet, remains true, the water around the Greenland Ice Sheet is bucking the trend. Worryingly, scientists fear that this cold spot is an early sign of pump failure.

What might be happening is this. As the difference between equatorial and Arctic sea temperatures diminishes, the northward flow of warm water from the equator slows. That's pump 1 beginning to fail. Furthermore, as more meltwater runs off the Greenland Ice Sheet (or less water is able to freeze), it dilutes the salty water. This restricts the maximum saltiness that can be achieved, which limits the maximum water density, thereby slowing the descent of water from the surface. That's pump 2 starting to fail. The combined failure of these two pumps means colder water around the Greenland Ice Sheet is not moved on as fast, leading to the observed cold spot.[224]

It has recently been suggested that the complete failure of this Atlantic system may occur much earlier than originally thought, well within this century. However, the experts do acknowledge that their predictions are unavoidably imprecise, owing to a phenomenon known as hysteresis.[225]

Flipping dangerously

It turns out that this chapter has already depicted hysteresis in action. It concerns those self-attacking immune cells that target the brain's communication networks in the most well-understood form of autoimmune encephalitis. Remember how those immune cells only produce their disease-causing antibodies after prior exposure to a specific neuronal antigen, most notably appearing within an ovarian tumour. This exposure shifts the immune cells from one stable state to another. Crucially, this new state is one that is primed to secrete the antibodies that cause the disease. What this represents is a form of cellular memory to a specific past exposure.

Systems that exhibit hysteresis, such as immune cells, are capable of remembering past exposures, because they act as analogue-to-digital converters. Put another way, they are very good at turning a scalar input into a polarised output. Faced with a variable signal, such as the quantity of a neuronal antigen in the ovary, there is a critical threshold that flips the immune cells from off to on. Importantly, once this occurs, attempts to reverse such a switch often fail.

Immune cells aside, the process of hysteresis is found in many natural systems. And the thermohaline circulation, the lifeblood of our oceans, is no exception. This is why scientists are struggling to chart its prospects accurately. Since it is so difficult to know where the threshold for flipping lies until *after* the flip, it is almost impossible to say if, or when, any early signs of pump slowdown will give way to failure.

This neat theory is all well and good, but why should we even care about the failure of these underwater pumps? One of the main reasons is that this process traps carbon dioxide from the atmosphere and shuttles it to the depths of the oceans. As it takes hundreds of years for the circuit to come full circle, carbon is locked away for some time. This has helped the ocean absorb around 30% of anthropogenic carbon release since the start of the Industrial Revolution, buffering the effects of climate change so far. However, pump failure is likely to be one of those self-perpetuating feedback loops we want to avoid, leading to retained carbon dioxide in the atmosphere and exacerbating the warming of our planet.

The slowdown, and possible failure, of these pumps will also have catastrophic effects on marine and coastal wildlife, as the temperature and nutrient content of the affected seas and oceans are altered. Biodiversity will inevitably suffer. Furthermore, changes to weather patterns are also predicted, mainly in the form of droughts and possibly colder temperatures in Western Europe, drastically reducing crop yields. Human welfare will inevitably suffer, too.

In this chapter, I have told just a part of our bipolar world's remarkable story. From our skies to our oceans, this is a story characterised by divergent extremes, flowing gradients and unpredictable flips. But somehow, despite showcasing such dynamism, I hope to have convinced you, too, of our planet's intrinsic fragility. Our world is repeatedly telling us how vulnerable it is, so why does this story so often get ignored?

The rule of binary

'There are 10 types of people in the world: those who understand binary, and those who don't.' I was recently reminded of this quip on a T-shirt advertised on my social media feed. And while I am at risk of shifting the numerical balance between these two types of people, its relevance to this discussion first necessitates an explanation. How does this phrase make sense?

The short answer is that **10** means '2' in binary notation. The longer answer involves a process of adding successive powers of 2 from right to left, beginning with 2^0. When there is a **1**, you add the corresponding power of 2. When there is a **0**, you add nothing. So,

$$1 \quad = \quad 2^0 \quad = \quad \text{One;}$$
$$11 \quad = \quad 2^1 + 2^0 \quad = \quad \text{Three;}$$
$$111 \quad = \quad 2^2 + 2^1 + 2^0 \quad = \quad \text{Seven;}$$
$$101 \quad = \quad 2^2 + 0 + 2^0 \quad = \quad \text{Five;}$$

And, of most relevance here:

$$10 \quad = \quad 2^1 + 0 \quad = \quad \text{Two.}$$

While binary is the universal language of computers, it is no stranger in the natural world. It underlies the hysteresis phenomenon I described in the previous section as a means for natural systems to remember past exposures. This concept is also integral to the all-or-nothing response of neuronal firing. Since many neurons transmit their signals over long distances, a scalar signal could easily become degraded by the time it reaches its destination, leading to inaccuracies. Instead, a binary signal maintains its message, making it a much more robust and faster method of communication. At a given moment in time, a neuron is simply either on or it is off.[226]

Binary owes its existence and utility to an endorsement of life's opposing extremes, represented by strings of 1s and 0s. North and south, love and hate, left and right, tails and heads, on and off, them and us, win and lose, evil and good, white and black, acceptance and denial ... it turns out that nearly all aspects of our cultures, livelihoods, surroundings, anatomy, beliefs and pastimes are infiltrated by this dualistic way of thinking.

But if we are not careful, we might become entirely defined by these extremes. Consider, for example, the sound wave and its two principal properties: loudness and pitch. While its loudness is determined by the differences between its peaks and troughs, its pitch relies on how quickly each of those extremes gives way to the other. Such a life ruled by extremes may be quite suitable for sound waves, but I would argue that it is not so favourable for the wider prosperity of Earth and its natural residents.

The reality is that even where binary systems exist in Nature, complementary mechanisms are also at play to facilitate a more graded output. Despite the nervous system's use of an all-or-nothing regime for signal transmission to the muscles, this does not constrain its host to some robotic dance of lurches and lunges. In health, muscular power is gradated, adaptable and coordinated. It allows humans to perform elegant waltzes, nightingales to sing their varied melodies, hummingbird hawkmoths to hover for nectar and cheetahs to outrun their next meal.

The way the brain exerts its control over this process is inevitably complex, gradually tweaked across successive generations according to the unique survival needs of each species. After millions of years of evolutionary fine-tuning, this control involves a combination of central predictions, sensory feedback and a strict hierarchy of exquisitely timed neuronal activation. The pool of motor neurons activating each muscle possesses a functional diversity that is critical. And the strength of signal by each neuron is communicated not by its size (after all, it can only be 0 or 1) but by the number of signals it transmits in a given time period.

My point is this. While dualistic systems play an essential role in the efficient functioning of the world, their dominance on a grander scale is not only unnatural, but wholly inadequate. In many contexts, I would not be the first to claim that they are actively harmful.

In *Not in God's Name*, Jonathan Sacks tells us that 'dualism is what happens when cognitive dissonance becomes unbearable, when the world as it is, is simply too unlike the world as we believed it ought to be'. Sacks goes on to argue that, in its most extreme, pathological form, dualism sees humanity 'divided into the unimpeachably good and the irredeemably bad', validating creeds that devote their existence to loving 'us' and hating 'them'. The suffix '-phobia' is commonly abused to scapegoat fear as the primary mechanism for such hatred. Furthermore, you only need to consider the prefixes 'anti-' and 'non-' to realise that the latest dualistic idea doesn't even require its own name these days. Dualism was not only born out of conflict but has grown up to become its most doting custodian.[227]

The duality of opposites

Wherever humans interact, opinions divide. At first, this may only concern the relatively inconsequential aspects of life, but wherever those divisions are allowed, if not willed, to polarise, the inevitable result is dualism. Humans, it seems, gravitate to the poles.

In order to explore this further, let me return to this book's central argument, which is that the challenge to conserve human health mandates an improved appreciation for all of life's natural processes.

Among those finishing this book, I, as the author, might be interested in the percentage of readers either agreeing or disagreeing with that core idea. This would give me a raw split between the two schools of thought. But to flesh that dichotomy out, I might want to know the proportion of readers liking or disliking each chapter. Evaluating it further still, I might choose to break it down, in ever greater painstaking detail, into sections, paragraphs, sentences, even individual words, thereby collating an ever longer chain of likes and dislikes (represented by 1s and 0s), each one representing an ever smaller fraction of the book.

Faced with such fine detail, I might reason along the following lines. To think that anyone finishing this book, even those agreeing with the central argument, might like every word in it would be ludicrous. To expect anyone to like every sentence would still be setting myself up for disappointment. The same would apply to the paragraphs and the sections, albeit to an increasingly smaller extent. Only when considering the split into chapters might it at last be likely (but not certain) that someone agreeing with the central argument of this book would also like each one of the resulting fragments.

On the flip side, I would be surprised if anyone in the disagreement camp disliked absolutely every word. While their chances of disliking every chapter would be much higher, this would not be guaranteed. I must say, though, that I would certainly commend such readers for finishing a book they disliked so strongly.

The conclusion I might draw from these rather elaborate musings is that a binary evaluation of the book's core argument imposes an artificial polarisation that is an extreme portrayal of its real, nuanced self. The upshot is that complex problems such as human health do not lend themselves to appraisal in a dualistic framework. Such an approach simply misses the innumerable greys between the black and the white that offer us the greatest tones.

A similar effect is observed when flipping a coin. After just one toss, the result (either heads or tails) is a one-sided representation of the coin's two-sided self. But the more the coin is tossed, we move further and further away from that binary misrepresentation, and the more balanced the string of results becomes.

Nevertheless, there is an important subtlety to this argument that I want to highlight. Polarised decisions on a smaller scale remain absolutely vital in the formulation of one's gradated, adaptable and coordinated viewpoint. This is just the same as the capability of all-or-nothing neuronal signals to produce our finely controlled muscular movements, also gradated, adaptable and coordinated. Perhaps this is no wonder, considering our brains, as the generators of our viewpoints, are a wired mesh of the same all-or-nothing neuronal gatekeepers that are involved in regulating our muscles too.

What we might deduce from this is that our brains are innately obsessed with duality, and, in particular, the duality of opposites. When my twin daughters were five or six, I remember enjoying a game of opposites over the dinner table from time to time. There seems to be something intrinsically mysterious and amusing about flipping the meaning of what we say. Indeed, a pathological form of binary reversal has also been observed in patients with a form of dementia that primarily affects speech output. No matter how hard those patients try, 'yes' often becomes 'no' and 'no' often becomes 'yes'.[228]

Whether or not this obsession with duality stems from our body and brain's anatomical split into two halves, I believe this speaks to our basic cognitive and emotional need to simplify our experiences in a world that seems set on perplexing and mystifying us. As Sacks pointed out, 'dualism resolves complexity'.[229]

Fuelling our imaginations

It turns out that this fascination with opposites is the perfect vehicle for imaginary thought. The real inspires the unreal. In the Hindu legend of Kanyakumari's origin that I outlined earlier in this chapter, the displaced gods concocted three items for Shiva to deliver, as the price for his virgin bride. These were based on removing the natural features of three common products of the region: the sugarcane stem with its rings, the betel leaf with its veins and the coconut with its eyes. In doing so, they delved into an imaginary world of *ringless* sugarcane stems, *veinless* betel leaves and *eyeless* coconuts.[230]

Without this innate tendency to fictionalise and embellish the world around us, there would be no stories of the richness and intrigue that most appeal. Ultimately, there would be no 'us', for our cognitive evolution has always been wrapped up tightly with our imaginative powers.[231]

But just like the two-faced façade of knowledge that I introduced in Chapter 2, with both its reputable and disreputable components, the human brain's affinity for altering the truth can also be a powerful disruptor. In modern times at least, emotionally charged soundbites on social media and punchy headlines among conventional press outlets are the playthings of fake news.

Faced with a streaming barrage of such information, the credible is so often swamped by the incredible that it becomes almost impossible to discern wisdom from bunkum. Instead, our brains choose between an echo chamber of familiarity or a firewall of exclusion, thereby either immersing themselves in what they already know or extracting themselves from the discomforting source of supranormal inputs. Either way, there is a loss of cognitive and intellectual diversity from the discourse, and opinions polarise. Ambiguity breeds conflict.

From a psychological point of view, polarisation of opinion is the product of cognitive traits, the social context and the interactions between them. Perhaps the trait that matters most of all is cognitive rigidity, which dictates how flexibly, based on an individual's genetic make-up and life exposures, they deal with change, conflict and closure.

Of course, what really matters is how an individual's traits play out among the rapidly changing social context of our technological and narrowcasted world, exposed to the rumours, rancour and rituals that characterise increasingly homogenised groups. Unfortunately, everything with a human brain is susceptible. If anyone considers themselves immune from this, no matter where their allegiances may lie, then it is likely they are among the most affected.[232]

But despair not, for there is good news. The most effective weapon in the psychological resistance against such division, should you wish to wield it, is your own insight. Once you recognise this dualism all around us, whatever the contested topic might be, you begin to see conflict for what it really represents: *a bipolar world in need of help.*

Protecting your identity

This brings me onto providing some sort of answer to a question I raised earlier in this chapter: why is the story of our fragile world being ignored by so many people? Instead of focusing, as we so often do, on those who either *accept* or *deny* climate change, worse still on those who *do* or *do not* believe in climate change, we need to diversify away from these polar extremes.

The climate scientist, Professor Katharine Hayhoe, is an advocate of this approach in her book *Saving Us: A Climate Scientist's Case for Hope and Healing in a Divided World*. It was while reading this down-to-earth book that I was introduced to Global Warming's Six Americas. Conducted jointly by the Yale Project on Climate Change and 4C of George Mason University, 2,129 American citizens were first asked about their views on global warming in the late 2000s. Six categories were formed based on the certainty, importance and anxiety that the respondents expressed in their answers: Alarmed, Concerned, Cautious, Disengaged, Doubtful and Dismissive.[233]

While only a quarter of respondents fell into the two extreme categories (18% Alarmed and 7% Dismissive), these two groups shared some interesting characteristics. Lo and behold, people in these two extreme groups not only felt they had the most knowledge on the subject (beware the towering façade), but they were also the most rigid in their thinking.

Seemingly, it was the groups in the middle, making up 75% of the respondents, who had the most to learn and were the most amenable to learning it. This was borne out when the latest round of the project in 2022 showed that since 2012 the Cautious group had shrunk by roughly the same percentage (–12%) as the Alarmed group had grown (+14%). All the other groups deviated by only 1–2%. Of course, what had actually happened was some of the Cautious had shifted into the Concerned group, and some of the Concerned had moved into the Alarmed category.[234]

What this all boils down to is *identity*. If you think and act as an Alarmist or as a Dismissive, you have the strongest sense of identity. As long as everyone around you conforms to the same thoughts and actions as you, you feel safe. But lurking constantly in the shadows is one of the biggest modifiers of any organism's behaviour. This thing ensconces us in bubbles. This thing is *threat*.

Let me explain what I am getting at here with a simple illustration from the natural world.

Standing sentinel on the underside of a silver birch leaf and shielding her brood of freshly laid eggs from predation and parasitisation, the Parent Bug *Elasmucha grisea* is a type of shield bug that truly lives up to her name. She goes on to protect her nymphs through all stages of their development, thereby sustaining the biological identity her predecessors evolved for her and ensuring safe passage for her progeny. Such instinctively maternal care is rare among insects and represents an extreme and resource-intensive way of dealing with widespread threat.[235]

Let's now contrast this overbearing behaviour with the underhand exploits of the Dark-edged Bee-fly *Bombylius major*. This bee-fly is the most common member of its genus and owes its reproductive success to a very different method of parenting, if indeed you can even call it that. Mimicking the solitary bees she targets in both looks and sounds, the female Dark-edged Bee-fly first arrives at the hole of an underground solitary bee nest. After covering each egg in dust and soil to aid the impending subterfuge, she flicks them individually towards the hole of the nest. The eggs that make it into the nest then hatch, and the larvae proceed to eat the host bee's grubs, before emerging as an adult Dark-edged Bee-fly. Of course, the female that originally tossed the eggs into the bee's nest is long gone.[236]

Just like the Parent Bug, the Dark-edged Bee-fly has evolved an identity that successfully manages safety and threat, albeit at the opposing extreme of parental styles. What would the shield bug think if it knew about the bee-fly's clandestine and exploitative methods? Similarly, how would the bee-fly act if it were suddenly told to show the same level of love and affection for its young as the shield bug? I strongly suspect they would be as alarmed and as dismissive as each other.

Asking the right questions

As I draw Part II to a close, let me summarise its takeaway themes. I began Chapter 4 by breaking down the duality of the seed and tree relationship, resolving their true unity as integral and integrated parts of one reproductive life cycle. I conveyed the RNA world theory as one of the most credible attempts so far at explaining the origin of life, a process reliant on both the fidelity of information and the efficiency of catalysis. I emphasised that the essence of biological stability is homeostasis and its self-rectifying feedback loops, while biological change ranges from the slow evolutionary march of environmental adaptations to the runaway self-perpetuating loops of systemic breakdown.

I took advantage of the conspicuousness of our own skin colour to exemplify the fundamental interactions between all organisms and their environments. In particular, I outlined how just the right amount of melanin in the skin of every one of our ancestors secured a healthy and fertile balance of life-sustaining nutrients, namely folate and Vitamin D. But without Nature's sunscreen blocking harmful ultraviolet rays and Nature's coolant regulating internal temperatures, we would have struggled to evolve the brainy, bipedal and bald simian niche we occupy today.

And when it comes to the planet we call home, I considered the critical role that plate tectonics, jet streams and underwater currents continue to play in regulating the Earth's global climate and regional weather patterns. The effects of colliding plates, jamming airflows and weakening oceanic pumps simply affect us all. Applying the same logic that we apply to the assessment of unwell patients in the hospital, I questioned whether the extreme symptoms and signs our transitioning planet is currently experiencing are akin to a regulated fever or uncontrollable hyperthermia. Furthermore, I hinted at whether this may be analogous to a self-destructive attack on our centres of communication and control. Importantly, I suggested that we have entered the climateric.

Then, inspired by the pervasiveness of Earth's polarity and the statistical inevitability of extreme events, I spent a lot of this chapter focusing on the duality of life. Whether we look to the Green Anole lizard switching its genes on or off, our neurons firing all-or-nothing signals to our muscles or the hysteresis of our immune cells, biology is built on binary systems. However, a bipolar world hell-bent on amplifying such dualistic thinking into all walks of life ignores the sublimity of variety, only serving to entrench harmful caricatures of an already fragmented society. And while lavish imitations of the true human experience allow our imaginations to run wild and healthy in an amphitheatre of stories and make-believe, the ease with which we turn the real into the unreal also gives way to deception, ambiguity and conflict.

Coming full circle and arriving back at the extreme parenting styles of the Parent Bug and the Dark-edged Bee-fly, it would be inadmissible of me to ask which parental strategy you instinctively prefer. It would most certainly contravene the whole message of this chapter if I were to ask you to choose your favourite between such contrasting examples of guardianship and abandonment. It would feel at complete odds to suggest such a decision might somehow influence our own custodial responsibilities – for this planet, for its natural inhabitants and for fellow humankind.

So, for all these reasons, I would never do such a thing. Instead, I would only reiterate how these would be precisely the wrong questions to ask. If I wrote this chapter to say just one thing, it is that our bipolar world deserves much more than a dualistic response to its ills.

In fact, the antidote to the plight of our natural world is simple. Luckily, this antidote is capable of exhibiting its beauty, humility, magnanimity and resilience in all corners of the globe. But it only possesses all these things when left sufficiently untainted.

The antidote is diversity.

Which is why I begin Part III by taking a much closer look at the biodiversity of Nature.

PART III

RECOGNISING NATURE AS A FORCE FOR GOOD

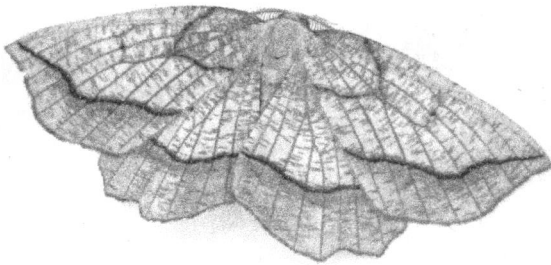

Bordered Beauty

CHAPTER 6

Biodiverse City

Life is what happens when chemistry competes

There is one for the coffee aficionado, several others for the nut lover and another set altogether for the one with a sweet tooth. There are those that distinguish meat-eaters from vegetarians and others that discern boozers from teetotallers. There are those superfluous to the vegan and those redundant to the Atkins dieter. There are those that cling on or move on, those that bunch up or split up, and some that only ever pass by, never opting to stop. They all make certain things, they all take certain things, but some do things no others can do. And this vast assortment lines a tunnel carved to extract the world's biochemical riches.[237]

This assortment is in fact the human gut microbiome, a staggering variety of microscopic organisms residing in the human digestive tract, at least as numerous as the 30 trillion mammalian cells that make up a human body. During my research for this chapter, I remembered that a friend of mine had recently requested a gut microbiome analysis from one of a host of commercial companies now offering this service. These companies will detect the microbial DNA in your poo, promising to reveal what the specific pattern of bugs says about your health. As I was intrigued to find out what such analysis could really glean, preferably without forking out the associated price tag, I was pleased when my friend agreed to send me his report.

First, this analysis outlines your overall gut diversity, placing you in one of four quartiles when compared to over 1,700 individuals from the general population. Second, the analysis focuses on the balance of 'good' and 'bad' gut microbes, supplementing their formal scientific names with more memorable aliases such as Otis and Cassie. It provides you with an individualised score, indicating your body's ability to cope with an influx of sugars and fats at mealtime. More specifically, it highlights how the 'good' bugs are associated with positive health parameters, such as higher levels of the favourable type of fat and lower sugar content in the bloodstream. Meanwhile, the presence of 'bad' bugs is generally associated with negative aspects of health, including raised blood pressure and greater abdominal fat. Finally, the report offers a customised list of foods to help shift the balance towards a more diverse and beneficial microbiome.

While I remain sceptical about how reliably one can apply even the clearest population-based associations at the individual level, what this exercise illustrated to me, beyond all else, is that *gut biodiversity matters*. But how and why, from an evolutionary perspective, should this have come about? I was beginning to think this might be a suitable parable concerning the biodiversity of the wider planetary ecosystem. After all, what is the gut if not an internalised pocket of the external biosphere?

How the gut came to be

To explore this, we need to step back approximately 600 million years, around the time that complex multicellular animal life was first appearing. It is worth clarifying that by this time Earth had already borne witness to the emergence of microbial, botanical and fungal lifeforms.[238]

As one newly fertilised egg divided into two cells, and two cells became four, then eight, sixteen and so on, the earliest animal ancestor shared its first moments of life, just as every animal still does today, as a lumpy clump of multiplicates. Further division would hollow out the centre, transforming that initial clump of cells into a fluid-filled ball with a cellular perimeter. But the simplicity of this spherical body plan put obvious restraints on the achievable complexity of primitive animal life.[239]

Since it is the external world that contains the primary ingredients for cellular health, the emergence of complex multicellular animals necessitated an efficient way of gathering and absorbing those ingredients. The breakthrough was just that: a pore that *broke through* the spherical body plan, thereby creating an internal cavity that was continuous with the nutrient- and energy-rich exterior. Crucially, this increased the surface area of the organism's environmental interface, permitting an enriched supply of water, sugars, fats and proteins to the cells demanding them. The primitive mouth had been born.[240]

However, it must be acknowledged that this was not the inevitable outcome for these early animals. Evolution could have gone the way of the multicellular fungus before it and sought interaction with the outside world in a different way. Instead of going in on itself, it might have gone outwards like the complex hyphal extensions of fungi today, constantly scanning, probing and extracting its immediate environment. Merlin Sheldrake, in his book *Entangled Life*, put it succinctly when he said that 'the difference between animals and fungi is simple: animals put food in their bodies, whereas fungi put their bodies in the food'.[241]

Nevertheless, the mouth has proven to be a very successful strategy for the animal kingdom. For some complex marine organisms alive today, such as corals and jellyfish, the mouth has satisfied all their digestive needs. Testament to the success of such a body plan, these animals rely completely on a blind-ended digestive pouch. But given that the way *in* is also the way *out*, the ingestion of nutrients and the egestion of waste must be sufficiently separated in time to ensure satisfactory absorption and avoid harmful contamination. The blind-ended digestive pouch, therefore, has its notable drawbacks.[242]

If you think back to Chapter 1, this draws parallels with the mammalian respiratory tract. Their cul-de-sac system of bidirectional airflow put limits on the mammalian anatomical blueprint when compared with their avian (and dinosaurian) competitors. In a similar way to the avian respiratory tract benefiting from a constant, unidirectional flow of air through their lungs, the primitive animal digestive tract was ready to take advantage of a similarly nifty upgrade.

It has to be said that experts do not fully agree on how this upgrade came about. One of the most credible demonstrations of what happened next comes from a type of ragworm, *Platynereis dumerilii*, widely resident today in European and Asian coastal waters. Considered a living fossil owing to its evolutionary stability over long periods of time, this small marine worm evinces a developmental history with far-reaching relevance to all animals.[243]

It turns out that one step better than having only one hole for exchanging nutrients and waste with the external world is to have *two*. Thus, one hole becomes the exclusive entrance to the digestive tract, while the other is the designated exit. In formal terms, this ragworm

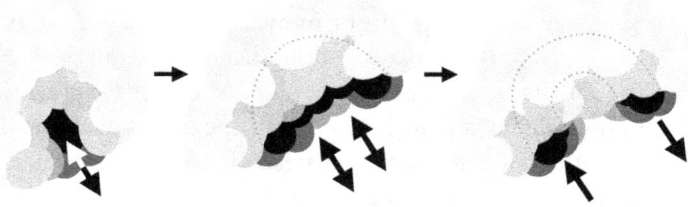

Figure 6. Primitive gut evolution in animals: how a single orifice for both food and waste became two, thereby permitting unidirectional conversion of food into waste.

is one of the earliest evolutionary examples of an organism with both a mouth and an anus, connected internally by an absorptive, one-way through-gut.[244]

Remarkably, these two holes originate simultaneously within the developing ragworm from the equivalent pore that forms the one-holed pouch found in corals and jellyfish. This process is complex, involving a pore that first elongates, slit-like, before closing up in the middle (Figure 6). But the very moment that the central closure is complete, one hole becomes two, joined internally by a continuous tube, known as the alimentary canal.

Just like that, the primitive animal gut had been born.

An axis of diversification

Akin to the seed–tree paradox at the start of Chapter 4, evidence suggests that the mouth and anus of most modern animals evolved simultaneously, thereby forming an unshakeable union that would go on to shape animal body plans in more ways than might initially appear obvious. Accordingly, this exposé of orifices would be incomplete if it did not address its uncanny association with the bipolar, binary theme of Chapter 5.

What this all revolves around is a bodily axis.

Central to the body plans of our worm-like ancestors, the alimentary canal formed an axis that defined left from right, head from bottom and front from back. It is difficult to summarise just how influential these fundamental distinctions have been right across the animal kingdom. The alimentary canal has both literally and metaphorically fuelled biodiversity.[245]

It has allowed a compendium of sensory organs to diversify around the mouth, perfectly calibrating the organism to its ambient conditions in the insatiable search for food. It has physically separated the ingestion of nutrients from the egestion of waste, which has maximised the time for absorption and minimised the chance of contamination. It has promoted a left–right symmetry that permits binocular vision, sound localisation and streamlined locomotion, while also providing a back-up in the case of one-sided organ loss. It has orientated organisms in the forwards direction, optimising linear movement when hunting for food, fleeing from predators and attracting mates.[246]

All these attributes, in the various guises we observe today, define a group of animals known as the bilaterians. In fact, the bilaterians include 99% of extant animals (including humans), only excluding a small selection of marine organisms such as sponges, jellies and corals. It follows that, since their divergence from a common ancestor hundreds of millions of years ago, the bilaterians have evolved guts of imaginative shapes, sizes and arrangements when compared with that of the ragworm.

Various parts of the alimentary canal have taken on specialist roles, from chewing to churning, from extracting to excreting, thereby providing ever greater differences between families of species. Many alimentary canals have gone one step further, merging with neighbouring bodily systems involved in breeding, peeing or breathing. Obvious examples of this include the role our own mouths play in how we breathe and the joint excretory system for fluid and solid waste from the cloaca of birds, reptiles and amphibians. There are examples of species with one mouth and many anuses, as well as species that devolved the need for mouth, gut and anus altogether. Even within the same species, guts transform according to the changing digestive requirements of complex life cycles.[247]

But despite these variations, the alimentary canal of most animals retains its basic structural form. It is essentially a tunnel with one entrance for food and one exit for waste. And healthy maintenance of this tunnel must combine regular replenishment from the natural world with timely elimination of any surplus.

Ultimately, it is these frequent exchanges with the outside world that carefully accrue, nurture and purify the band of microbes living within those cylindrical walls. As my friend's gut microbiome report confirmed, it is clear a gut's life is overwhelmingly a bug's life.

But what happens when it isn't?

Ever since the first experiment of its kind was performed in Germany and published in 1897, the rearing of bug-free animals has helped answer this question. The gut walls of these germless animals are thinner, less varied and slower to regenerate. Having fewer intrinsic nerve endings hampers food transit, and the presence of fewer immune cells impairs host defence. The animal becomes more reliant on dietary sources of key vitamins, and the first part of the large bowel distends by as much as tenfold. Even outside the gut, the blood–brain barrier becomes leakier, the stress response is amplified and sociable behaviour is diminished. It turns out that the bugless life is a *nicht-so-gut* life.[248]

Given the primacy of the gut and its resident microbiome in animal development, perhaps these widespread effects should come as little surprise. After all, biological systems generally cope poorly with loss. If we extrapolate these findings to our own species, it suggests the bugs in our guts have been well placed to guide human evolution right from the origin of animals 600 million years ago. And no organ validates this assertion more than the brain.[249]

How the gut feeds the brain

Stooped and static, my next patient stops abruptly at the jambs of the door. 'Please take your time,' I say, as his frozen posture soon thaws, and he festinates, one hand with stick, the other stuck mid-swing, towards the clinic room chair. He swivels awkwardly, sits and faces me, staring, almost facelessly. He is thin and frail, natural sequelae of a shaky swallow feeding a sluggish stomach. His right hand refuses to rest, shifting rhythmically back and forth as though caught, unwillingly, between two opposing worlds.

These are the classical neurological features of Parkinson's disease, resulting from the depletion of a key chemical (dopamine) in a part of the brain critical for the initiation of bodily movements. It may be the second most common degenerative disease of the central nervous system after Alzheimer's dementia, but recent evidence has forced scientists in search of its cause to think outside the brainbox.[250]

Intriguingly, this disease begins decades before patients walk through the doorway of a neurology clinic. In many patients, these prodromal years give rise to constipation, poor smell and a specific sleep disorder whereby individuals physically act out their dreams. And wherever scientists look to explain these early symptoms, the same protein crops up, clogging up cells. Notably, this finding is not exclusive to the brain but extends to cells in the gut, too. This protein is alpha-synuclein, the same one found in the brains of patients with established Parkinson's disease.[251]

So what is going on? Could the gut really be involved in the development of Parkinson's disease, even though it is traditionally considered a disease of the brain? Could remote events in the gut even be the cause of this 'neurological' disease? If so, then surely an analysis of the gut's resident microbes would provide critical insight. Indeed, that is exactly what was done.

Applying a similar method to the one used to formulate my friend's gut microbiome report, one group of scientists impressively looked at the microbial differences between 490 patients with Parkinson's disease and 234 individuals without. Of the 257 identified microbes that passed their stringent quality control, a third were either enriched or depleted in patients with Parkinson's disease. And the differences were not trivial. For microbial species at the most extreme ends of the spectrum, there was an approximate sevenfold difference in abundance between the two groups.[252]

When the researchers looked at these specific microbial patterns, what became clear was the vast array of mechanisms by which the gut microbiome can influence the brain. Generally, for anything we ingest to harm the brain, it must first interact with our resident gut microbes. In this way, our gut bacteria are the first line of defence against any potential attack from within, however insidious the resulting harm may turn out to be.[253]

Our gut bacteria control the permeability and structural integrity of the gut wall, and they superintend the immune cells primed for activation. They modify the tablets that doctors prescribe us, and they disarm the toxins that are inadvertently served to us in our food. While all these factors dictate the baseline absorption rates of potentially brain-bending substances present in our diets, our gut bacteria actively contribute much more besides. They not only make important vitamins upon which neurons in the brain rely, but they also produce the same chemicals with which neurons in the brain communicate. In addition, they ferment fibre and metabolise plant-based sugars, converting these common dietary constituents into key anti-inflammatory, pro-motile mediators within the gut wall.[254]

In Parkinson's disease, the combined enrichment of some gut bacteria and depletion of others is hypothesised to undermine each one of these mechanisms in some way. But, hang on, isn't there a missing link here? How can these local changes in the mid-gut feasibly affect the brain up near the mouth end?

While the bloodstream is an essential transport route from the gut to the brain, first detouring via the liver for further chemical sifting, a direct neuronal express is also in operation. Despite the physical separation between the brain and the majority of the alimentary canal, it turns out that the vagus nerve (the tenth cranial nerve) connects the two in a bidirectional manner. Almost certainly a legacy of the brain's inescapable reliance on the gut from the earliest stages of animal evolution, the vagus nerve is an integral part of what is widely referred to today as the microbiome–gut–brain axis.[255]

Remarkably, in post-mortem studies of patients with Parkinson's disease, the vagus nerve contains the very same alpha-synuclein protein clogging up its cells as the one found in the patients' brains and gut walls. Braak's hypothesis is famous within this area of research as a way of reconciling these findings. According to Braak, Parkinson's disease originates in the microbial gut (and nasal cavities), before ascending over many years via the vagus nerve to the brain. It is only then that it finally reveals itself in every undeniable bit of its stooped, festinant, tremulous and faceless form.[256]

And with that, we have spun full circle. From the primitive alimentary axis on display in marine ragworms to the highly adapted microbiome–gut–brain axis fundamental to the human body plan, I have outlined how new knowledge in this field continues to revolutionise recently held conventions.

But one key aspect of my message still puzzled me. Despite my initial instinct and onward insistence that *biodiversity matters*, studies so far have concluded that gut microbial biodiversity is not actually reduced in Parkinson's disease. The balance between different groups of microorganisms undoubtedly shifts, but overall biodiversity is unchanged. Faced with this fact, it hardly serves as the worldly parable for Nature's plight I thought it might. Had I got it wrong? Is biodiversity itself not that important after all? Or does this simply reflect the fact that all useful analogies eventually fail? While I am always happy to admit my errors, I am also not one to give up easily. As such, I felt compelled to first delve more deeply into the very essence of what biodiversity really is.[257]

Nightly visitors in their droves

In the garden, I have seen tigers dressed in scarlet and ruby. I have witnessed leopards, elephants and mice, too. I have glimpsed antlers, ears and snouts among flashes of copper, crimson and blue. I have had pearls and emeralds laid out on carpets of various types. I have been visited by peacocks, swifts, magpies and the like. Old ladies and footmen have all trodden softly amid the poplars, toadflax and box. I have seen many varieties of pugs and hawks, and on one occasion, even a fluffy female fox.

Since April 2022, the deployment of ultraviolet bulbs at night has attracted over 250 new species to my small urban garden in Essex, UK, plus another 150 further afield. Both their colloquial and scientific names are as diverse as the colours, shapes and behaviours that identify them. Each species is finely attuned to the whims of its favoured season, and their numbers from night to night vary according to the prevailing winds and the intensity of moonshine. Yet they are largely shunned, or at best forgotten, by the many who perhaps fear, or at least misinterpret, their night-time proclivities.

All the species I allude to are part of an ancient order of insects dating back 300 million years. They have seen the rise of the dinosaurs followed by the fall of the non-avian ones. They have clung on through mass deaths and meltdowns, only to flourish as the perfect accomplices for flowering plants and trees in the aftermath.

The diverse group of species I describe are the moths. But if, like me, you prefer a literal translation that exists among the romantic European languages, then you can also think of them as the butterflies of the night.

Although the evolutionary branchpoint of butterflies from moths goes back about 100 million years, it is difficult to justify the favoured status that butterflies instinctively hold among humans today. Undoubtedly, this has more to do with our shared preference for the daytime than anything else. Because of this, butterflies are widely considered symbols of the bright, bold and beautiful, while moths have unjustly become icons of the dull, dark and dusty.[258]

All told, there are at least 160,000 known species of moths and butterflies worldwide, with estimates of another few hundred thousand yet undiscovered. In terms of species numbers, moths outnumber butterflies by at least seven times globally, increasing to forty times in temperate climates like the UK. While you would be hard pushed to locate the smallest micromoths with the naked eye, only microscopic examination of lock-and-key genitalia distinguishes larger species that look identical otherwise. This degree of biodiversity, playing out on so many levels, is what truly staggers me. And that is without even mentioning the huge variety of caterpillars that give rise to the flying adult forms.[259]

On numerous occasions I have been asked what drives my nascent interest in moths. Is it the remarkable complexity of their life cycles? Is it their mesmerising wing patterns? Is it the longevity of their evolutionary tale? Is it their intricate symbioses with plants and trees? Is it their surprising docility during the morning inspection? Or is it their almost universal accessibility wherever in the world you are?

While my answer often includes some variation of these qualities, there is also an element that is none of these things at all. Steadfast yet mysterious, there is something intangible that glues everything together, a connection that refuses to be pigeon-holed. Like many others before me, I have begun to appreciate moths as the perfect doorstep portal into the expanse, wonder and vitality of Nature.

In celebration of these remarkable insects, the Natural History Museum in London, UK, houses 12.5 million specimens of moths and butterflies in a large cocoon in the middle of its Darwin centre. While the pinning of specimens has thankfully been replaced by digital photography (and subsequent release of the unharmed insect) among amateurs like myself, it remains an historically important way for scientists to catalogue the world's lepidopteran riches. In June 2023, my wife arranged for us to meet with the Senior Curator in Charge, Geoff Martin, for a behind-the-scenes, by-appointment-only tour of the museum's 80,000 drawers.

Given the sheer breadth of the collection, Geoff had advised us to come with a plan. Did we want to see natives from Asia, Africa or the Americas? Were we interested in the big or the small? Was it bold colours or cryptic patterns we were most fascinated by?

As the spectacular Emperor Moth *Saturnia pavonia* had recently visited our garden in Essex, I first asked to see another marvel from the same family, the Atlas Moth *Attacus atlas*. With a wingspan of up to 24 cm, this palm-dwarfing Asian moth gives the illusion of being only half its size, ominously flanked by two deterrent serpentine heads. Also in the Asian section, Geoff showed us the scintillating splendour of a whole drawer of Owl Moths *Brahmaea wallichii*, whose tightly packed wavy lines serve to hypnotise any onlooker. Equally mesmerising was the technicolour display of the Madagascan Sunset Moth *Chrysiridia rhipheus*.

Geoff then showed us the Uruguayan Palm Borer *Paysandisia archon*, a large brown moth hiding a trio of orange, black and white nestled in its underwing. He taught us how this moth has inadvertently made its way to southern Europe via modern transport networks, where it has ravaged palm trees in the same way Asian Elm Zigzag Sawflies have devastated elms in North America. He also took us to a drawer of African Death's-head Hawkmoths *Acherontia atropos*, a black and yellow, somewhat sinister-looking moth, owing to the appearance of a human skull on top of its head.

Finally, we completed the tour with a comparison of two common UK species that are far less visually impressive. But what those drawers, replete with the almost indistinguishable Grey Pugs *Eupithecia subfuscata* and Common Pugs *E. vulgata*, made me reconsider was something that had always seemed so clear. And yet, when I really thought about it, that mask of clarity shattered. Akin to the study of DNA without knowledge of its constituent bases, it dawned on me that I could never understand biodiversity without a better definition of *its* basic unit. I needed to explore the origin of *a* species.[260]

But before I elaborate on how new species come about, we must first define the term. According to one definition, species are 'groups of actually or potentially interbreeding populations, which are reproductively isolated from other such groups'. That seems simple enough, you might think. But the concept of what constitutes a species is not always straightforward. Let me now share a particularly intriguing example.[261]

An interbreeding paradox

Encircling the flats of the Californian Central Valley is a mountainous ring that harbours a special power. Like an eye capable of looking into the past, this geographical arrangement plays storyboard to an evolutionary epic. The starring role is a species of salamander known as the Ensatina *Ensatina eschscholtzii*. And it remains one of the finest examples of what scientists call a ring species.[262]

Considering the word 'species' stems from the Latin 'specere', meaning 'to look', it seems apt to begin with a description of this salamander's looks. The inland populations have a blotched pattern, ranging from red blotches on brown to yellow blotches on black. For members of one population, their brown and beige blotches have even coalesced into a pattern closer to stripes. Meanwhile, among the unblotched coastal populations, individuals display a variety of tonal gradations from the animal's back to its front. Based on looks alone, you might be hesitant to classify all these groups as the same species.

But looks barely scratch the surface. Despite their varied appearances, these salamanders share a set of well-honed, survival-enhancing characteristics. For instance, all of them are lungless, nocturnal and capable of regenerating their detachable tails. Sharing such functional attributes, it is hardly surprising that neighbouring populations successfully interbreed. Well, most do. But, of crucial relevance to their classification as a ring species, not all of them do.

In order to explain why all pairs of neighbouring salamander populations from this species successfully interbreed *except one*, we must go back and look at the role the Central Valley has played in the salamander's evolution.

It turns out that the Central Valley is not a very welcoming environment for this salamander. This species is much better suited to the surrounding mountains, and therefore the salamander has been forced around this inhospitable void as it has descended from the north. In effect, this species has been split into two, thereby creating the blotched inland group and the unblotched coastal group.

As these two evolutionary paths advanced separately over countless generations, distinct populations established themselves, giving rise to their variations in appearance. Crucially, each population retained sufficient genetic similarity with its neighbours to interbreed success-fully. This maintained the genetic flow and kept them conjoined as one species.

But the Central Valley does not continue indefinitely, and eventually the two advancing groups of salamanders met in the mountains that mark the valley's southern limit. Remarkably, those two divergent populations had become genetically dissimilar enough that sustainable interbreeding between them was not possible.[263]

Herein lies the apparent paradox. Despite the ring of salamanders demonstrating the genetic continuity of a single species, the lack of interbreeding between the extreme populations contradicts such classification. Indeed, it conflicts with our original definition of a species.

So how is this paradox resolved? One approach is to realise that it never was one in the first place. Instead, what we are observing is a transition between two states: the first state contains one species and the second state contains two. The transition will complete when a second break in genetic flow somewhere along the ring of salamanders finally splits the current species into two. This could happen if one or more of the interbreeding populations were to become extinct, or if a new geographical barrier were to prevent contact between neighbours. Essentially, a ring species is a snapshot of species formation in mid-flow. It evades neat classification, because Nature cares little for the organisational demands that the left sides of our brains impose upon us.

Despite this, I do acknowledge that robust classification *is* firmly rooted in science, and so it needs to be. There is no better example of our dependence on it than our continued use of Carl Linnaeus' eighteenth-century binomial naming system for all species today. However, an acceptance of its imperfections continues to inspire revisions, especially as modern genetic analysis usurps a system that was primarily based on the looks of species.[264]

Whether or not the salamander ring species that circumscribes California's Central Valley will ever be reclassified as two (or more) distinct species, only time will tell. Like a stopwatch recording the mile splits in a marathon, this ring symbolises both the continuity of evolutionary flow as well as its propensity for artificial division.

In accepting these subtleties, I now appreciate that the concept of a species is not as basic as I instinctively thought. Charles Darwin recognised this with his description of the now famed Galápagos 'finches', which showed gradations in beak shape and size that correlated with a spectrum of seed preferences. Indeed, it was observations like this that drove him to propose his elegant theory of natural selection in 1859. Scientists today recognise this process in the intriguing and paradoxical nature of ring species. And how this is one way new species can be formed.[265]

But what I still didn't know was how our adaptable classification of species is translated into measurements of biodiversity. Surely, I needed that knowledge to decide whether biodiversity truly matters or not. Surely, this was essential to decide whether biodiversity is a genuine conserver of health.

A simple measure of biodiversity

Imagine I have a black velvet bag filled with 12 red balls. If, without looking, I were to pick one ball at random, replace it into the bag and then pick another, what is the chance that both selected balls would be the same colour?

The answer, of course, is 100%, or 1.

But then imagine that I remove half the red balls and I replace them with the same number of blue balls. So now I have the same black velvet bag, but this time it contains six red balls and six blue balls. Now if I were to pick one ball, replace it and then pick another, what is the chance both balls would be the same colour?

The chance of both balls being red is ½ (50% chance of the first ball being red) multiplied by ½ (50% chance of the second ball being red). This can be simplified to $(½)^2$, which equals ¼. The same is true for the probability of two blue balls being picked, so the answer is these two probabilities added together: ¼ (chance of two red balls) + ¼ (chance of two blue balls) = ½.

Extending this further, imagine I then repopulated the black velvet bag so that it contained four red balls, four blue balls and four green balls. Now if I posed the same question and applied the same logic, the answer would be ⅑ + ⅑ + ⅑ = ⅓.

For three red balls, three blue balls, three green balls and three yellow balls in the bag, the answer is ¼. For twelve balls of all different colours, the answer is 1/12. I am sure you can see the pattern.

So far, though, this does nothing more than prove what is blatantly obvious; as the variety of ball colours increases, the chance of finding similarity decreases. But let's flip it around.

What if, instead, I wanted to calculate the chance of picking two *different* colours in each of our experiments. All we would need to do is subtract each of the probabilities from 1. In doing so, we have just devised a method for calculating diversity. As the variety of ball colours increases, the chance of finding difference increases too.[266]

It turns out that this simple measure of diversity requires two types of input. We need to know how many different colours there are in total, which is the so-called richness of the balls in the bag. But we also need to know the relative proportions of each colour. While I have explored the effect of increasing the richness of the sample, I have not yet altered the relative proportions.

Take the two-colour experiment described earlier, and let's instead put eight red balls and four blue balls into the bag. The proportion of red balls is ⅔ and the proportion of blue balls is ⅓. Now when we calculate the chance of picking two different colour balls, we end up with $1 - ((⅔)^2 + (⅓)^2) = 1 - (4/9 + 1/9) = 4/9 \approx 0.44$. This probability is lower than the one calculated in the evenly distributed scenario, which is $1 - ½ = 0.5$. Crucially, it turns out that any deviation from the most evenly distributed scenario reduces the chance of picking different colour balls. This holds true no matter what the richness of the population may be. Evenness promotes diversity, while uniformity damages it.

When we extrapolate this to the ecological measure of biodiversity, we are therefore interested in both the richness and the evenness of a population. The richness is how many species there are and the evenness reflects how similar their relative proportions are. What the maths and our intuition agree on is this: the richer and evener a population is, the more biodiverse it becomes.[267]

A stimulating simulation

In order to test this further on my own terms, I set up a simulation using the same programming software I have been using throughout my research. I started by assimilating a digital population of 100 individuals, all from the same species. You can imagine this species as a bacterium, a plant, a moth or any living being for that matter. I pretended it was the Ensatina salamander from the Californian mountains.

The next thing I factored into the model was that this species, and all subsequent species generated by the computer's 1s and 0s, had both a speciation rate and an extinction rate. In other words, as each unit of simulated time passed, every species had a chance of splitting into two distinct species, and each one had a chance of dying out.

Importantly, I programmed the speciation rate to be proportional to the size of the species' population. I figured that the more individuals there are within a given species, the greater the likelihood that a species-forming set of mutations might manifest. In contrast, I presumed that the rate of extinction was *inversely* proportional to the species' population size. Here, I was assuming that a potential extinction event was more likely to threaten a less abundant species than a more abundant one.

After each arbitrary unit of time, I instructed the model to calculate the biodiversity of the population once any speciation and extinction events had taken place. To do this, I used the same method for calculating biodiversity as I applied to the balls in a bag. That way, the combined impact of adding and removing species in this virtual ecosystem could be tracked.

Given that the initial population consisted of only one species, the biodiversity started at 0. After all, this is identical to the chance of picking two different colours from a bag full of red balls. Then, having set the speciation rate so that it exceeded the extinction rate, biodiversity increased rapidly in the simulation before settling at just over 0.9 (Figure 7A). Next, I looked at the impact of changing the extinction rate while holding the speciation rate steady. Unsurprisingly, the higher I set the extinction rate, the more volatile the biodiversity became. I was satisfied that this simplistic model was behaving as I expected it to.

But what I was most interested in was not how one species becomes many, but instead how many species can be conserved. More specifically, I wanted to know how an established population might respond to changing environmental pressures. And now that I knew biodiversity was influenced by both richness and evenness, I was curious about their relative contributions in promoting ecosystem resilience.

To tackle this, I next manipulated the starting population of 100 individuals in two ways. My idea was to create two distinct populations with the same biodiversity. One population would achieve this by having a high richness of 20 species, while the other would have a low richness of only 10 species. In order for these two populations to compute the same biodiversity of 0.9, the evenness had to differ. So, in the population with high richness, a low evenness was required. And in the population with low richness, a high evenness was needed.

Through trial and error, I separately identified which set of opening simulation parameters produced a stable ecosystem equal to the biodiversity of my two starting populations (=0.9). This happened to contain 12 species with a moderately high level of evenness. By then

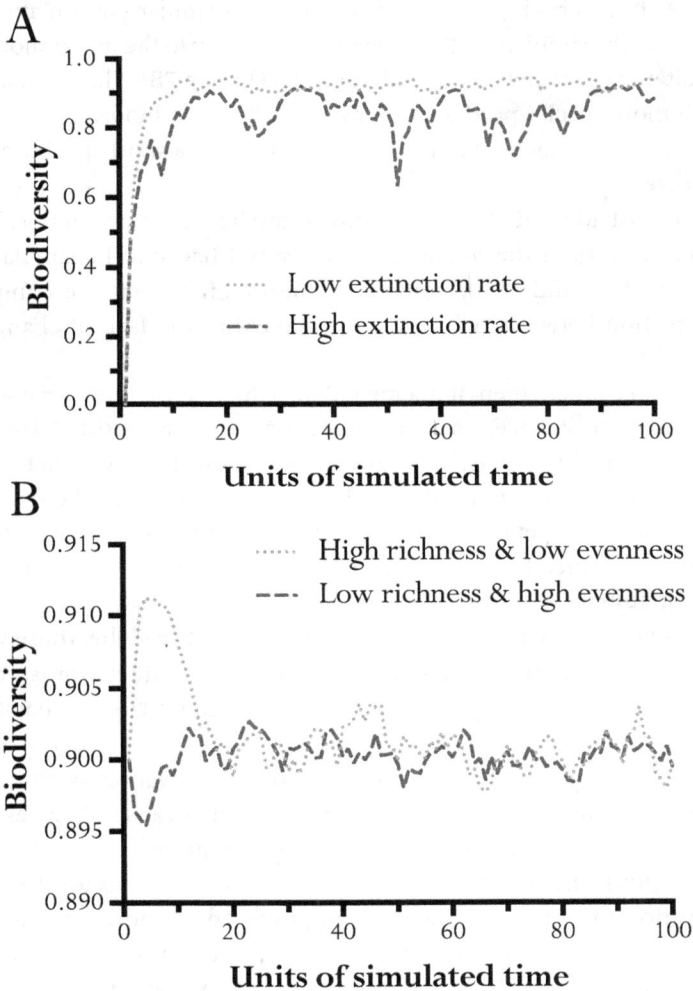

Figure 7. Biodiversity simulation: A. Effect of a variable extinction rate on ecosystem biodiversity while the speciation rate remains unchanged. B. Relative contributions of richness and evenness to ecosystem resilience in the face of new environmental pressures.

exposing my two 'extreme' populations to this particular model, I was aiming to isolate the relative importance of richness and evenness. In essence, I was simulating how each population would react to a changed world.

Although both populations settled into a similar pattern after 20 cycles of the simulation, they behaved differently to the initial shock of an altered set of environmental pressures (Figure 7B). The population with more species produced a temporary boost in biodiversity, while the species-depleted population instead suffered a short-lived drop in biodiversity.

I must admit that I wasn't entirely sure how to interpret this. Was it just a quirk of the rough and ready way I had put the simulation together? Or did it say something meaningful about the complex interaction between richness and evenness in the face of changing survival pressures?

If the latter, then it seemed that a higher richness promoted ecosystem resilience, as biodiversity was initially boosted for the population with high richness and low evenness. However, that boost in biodiversity was not brought about by an increase in the number of species. It was entirely due to the population becoming more even. Could it be, therefore, that an ecosystem is most resilient when it has a high number of uneven species?

When real ecosystems are examined across the globe, those with the greatest richness are also the most uneven. In other words, these rich ecosystems, which tend to be found in more tropical habitats, contain a minority of common species and a majority of rarer species. An intriguing question arising from this observation is this. Do ecosystems naturally favour this arrangement because it acts as the best shock-absorber to environmental perturbations?[268]

If this is the case, then it has significant implications for the conservation of our ecosystems moving forward. It indicates that the rare species are just as important as the common ones. Even more than this, it suggests that the as yet undiscovered species, which are likely to be heavily biased towards the rarer species, are just as crucial to the planet's ecosystem as the already discovered ones. This makes the tragedy of Centinelan extinction, which is the permanent loss of a species before we even discover its existence, all the more poignant.[269]

Seeking some real-life evidence

At this stage, I felt that I had exhausted the utility of my oversimplified computer model. And so I decided to explore the many biodiversity simulations devised by the experts. This quickly reinforced the inherent complexity of our planetary ecosystems.

There aren't just speciation and extinction to account for, but there are also the adaptations, interactions and movements of species to factor in. Instead of simply counting the number of individuals in an ecosystem, one also needs to model colonisation patterns, genetic markers and evolutionary pathways. This is before you even try to identify the impact of invasive species, climate change, human exploitation and natural geological processes. To simulate biodiversity, it turns out, is to simulate Nature itself.[270]

This reminded me of something from Chapter 2. As useful as models are in generating and assessing hypotheses, they are always an imperfect imprint of reality. Their core basis in the theoretical will always constrain their true bearing on the natural. To advance my understanding of biodiversity, I was in need of some real-life evidence.

This arrived in the form of a remarkable project that has been running since 2002. Focusing on the performance of a grassland ecosystem, the Jena Experiment arguably provides the most important insight so far into the true impact of biodiversity.[271]

Just outside the city of Jena in Germany, an area the size of 12 football pitches has been systematically divided into over 400 plots. Each one has either been sown with between 1 and 60 species of plant, or it has been left as a control plot. From day 1, scientists have been collating a vast dataset, comprising measures of plant growth, ecosystem health, water balance and nutrient cycles.

In a summary of the first 85,000 measurements spanning 15 years, the benefits of biodiversity were plain to see. First, plots with a higher richness of plant species produced taller plants with bigger leaves and more extensive root networks, together augmenting the overall yield of plant biomass. Second, plots with a higher richness of plant species produced a higher richness of animals, microbes and fungi both above and below the ground. Third, plots with a higher richness of plant species stored more carbon and made more efficient use of rainfall. Importantly, several of these correlations only became apparent after a time lag. For example, it took four years for the positive relationships with microbial biomass and root networks to materialise.[272]

The beneficial effect of biodiversity on plant yield was so strong that it outcompeted the effects of fertilisers and mowing intensification in artificially managed subplots. Crucially, biodiverse plots achieved this without any of the ecologic and economic costs associated with these intensified management practices. It turns out that biodiverse ecosystems host the most tireless of toilers.

There is no doubt that the Jena Experiment has proven biodiversity's value in a grassland ecosystem. But to which members of the ecosystem is biodiversity most valuable? After all, shouldn't we be clear about the natural beneficiaries of biodiversity?

The plants benefit, as they can perform more photosynthesis given their larger, more elevated leaves supplied with plentiful water from their extended root networks. The insects benefit, as they have greater access to the oxygen, sugars and shelter produced by the plants. The microbes benefit, as they fulfil their metabolic demands in recycling carbon, nitrogen and oxygen through the productive ecosystem. The fungi benefit, as they decompose a healthy abundance of dead plant material undersoil. And together, they all benefit, as they individually have a greater chance of successful reproduction.

Ultimately, a biodiverse ecosystem bestows these benefits on its occupants in the same way that organisms adapt their own internal performance for survival. Ecosystems, like organisms, evolve, since it is the most resilient among them that survive. But this all hinges on one fundamental concept. Within successful ecosystems there must be a varied selection of species, and within successful species there needs to be plenty of variation, too. We need no greater evidence of that than the Ensatina salamanders from California. From outside to in, there is variation in the looks, in the guts and in the genes of every species we might care to explore. From inside to out, it is biodiversity that expresses and amplifies itself through all the glorious layers of the natural world.

What could be more important to conserve than that?

The language of Nature

Biodiversity is fundamental to human well-being, a healthy planet and economic prosperity for all people. We depend on it for food, medicine, energy, clean air and water, security from natural disasters as well as recreation and cultural inspiration, and it supports all systems of life on Earth.

Signed by 196 nations in December 2022, these are words taken from the *Kunming-Montreal Global Biodiversity Framework*, which vows to conserve at least 30% of the planet for Nature. This would contribute to its 2050 vision of a world where 'biodiversity is valued, conserved, restored, and widely used'.[273]

This is a grand plan stated in no uncertain terms.

And given this chapter's preamble into the nuances of species classification, the fundamentals of biodiversity measurement and the mechanisms of ecosystem function, we are better equipped to understand how such a plan will help to conserve the health of the natural world. But what is perhaps less clear is why any of this is even necessary.

While writing this chapter in September 2023, the UK's fourth *State of Nature* report was released. The UK may only make up 0.05% of the Earth's surface, but its island geography, relatively stable climate and predilection for chalk streams have combined to make it an impressively biodiverse part of the world. Sadly, as this 2023 report confirms, this natural legacy only survives in showing how far the mighty can fall.[274]

Since 1970, the overall abundance of terrestrial and freshwater species is down by a fifth, while the abundance of UK priority species is down by two-thirds. Of most concern, 1 in 50 of UK species is now extinct and 1 in 6 is at risk of the same fate. But that's not all.

Devised by a team of researchers at London's Natural History Museum, the Biodiversity Intactness Index estimates the long-term impact of human activity on ecosystem richness. It does this by calculating the percentage of pre-modern species that have survived in a given locale. According to this analysis, the UK's Biodiversity Intactness Index is an abysmal 53%, way below the global target of 90%, cementing the UK as one of the most Nature-depleted countries in the world. Even at the global level, an average Biodiversity Intactness Index of 75% asserts the widespread need for much greater protection of our ecosystems.[275]

But while the lessons from the UK are undoubtedly stark, one should not discount some faint glimmers of hope. The decline in UK priority species since 1970 has at least plateaued in recent years, and some species, particularly wintering waterbirds, are showing increased numbers over the long-term. In addition, there is now higher expenditure on biodiversity, greater levels of volunteering for Nature and wider areas of sustainably managed land than two decades ago, all of which could help to turn the tide.[276]

As vital as these reports are, I must admit that it is easy to get lost amid the multitude of data, graphs and percentages. It is simply not an instinctive function of the human brain to weigh up this sort of information. While this skill can certainly be learned and honed, even then it remains somewhat abstract and remote from our lived experiences. Undoubtedly, this continues to undermine how our planet's loss of biodiversity is collectively perceived. The streams of data leave many people with an overwhelming sense of 'so what?'

Accordingly, I have been grappling with a suitable illustration of what's really going on, something that our brains do instinctively resonate with. As I gazed at the pages of the UK's *State of Nature* report, I finally realised that the answer was staring back at me the whole time. I decided to conceptualise the biodiversity problem using a core aspect of our cognitive development: language.

'I whether will be to the meaning this if remove other.' If you are as confused as I am on reading this sentence, then my suspicion is confirmed. After all, the sentence should read, 'I wonder whether anyone will still be able to understand the full meaning of this sentence if I remove every other word.'

If language is an ecosystem, then its words are its species, and its letters are its genes. The distinction between verbs and nouns becomes as profound as the difference between animals and plants. Its grammar outlines how species interact, and their written, spoken and signed expressions mediate their reproductive success. Abbreviations and acronyms are types of mutations, and synonyms are its insurance policy. Long-lost words finally become extinct, as newly coined words emerge. Homonyms mimic and homophones imitate. And languages intermingle in the same way that ecosystems overlap.

Ultimately, it is a language's vocabulary that represents an ecosystem's biodiversity. Since the function of language is communication, then a constricted vocabulary reduces its ability to convey the full richness of meaning. A language without words becomes meaningless.

In *The Lost Words*, Robert Macfarlane and Jackie Morris grieve the most telling of cultural sorrows:

> Once upon a time, words began to vanish from
> the language of children. They disappeared so
> quietly that at first almost no one noticed – fading

away like water on stone. The words were those
that children used to name the natural world:
acorn, adder, bluebell, bramble, conker – gone!
Fern, heather, kingfisher, otter, raven, willow,
wren… all of them gone![277]

As a father, I am proud that my daughters take delight in reciting the beautifully illustrated poems within this book. This is vital if we are to revive this vanishing lexicon of life.

In drawing this chapter to a close, it is only right that I now reflect on the quest I set myself earlier in this chapter. Am I any closer to knowing what biodiversity really is?

Well, I suppose biodiversity can mean just about anything you choose. Its very essence is the variety that means a catalogue of things to all different people.

To the mathematician, it is sets of equations.

To the lepidopterist, it is drawers of specimens.

To the ecologist, it is plots of grasses.

To the linguist, it is phrases of prose.

And to any one of the countless species occupying niche upon niche across ecosystems the world over, biodiversity is the language of Nature.

CHAPTER 7

The Benevolence of Nature

On stamping our mark within the Symbiocene

I wonder how you would rate your own relationship with Nature. One simple gauge (Figure 8) requires just two circles, whereby one represents 'self' and the other symbolises 'Nature'. When you think about your own connection with Nature, how much do these two circles overlap? Which out of A–G would you choose?

This seven-point evaluation ranging from complete detachment to complete immersion is known as the Inclusion of Nature in Self scale. And it is one of the simplest ways to measure so-called Nature connectedness. Yet, despite its ease, it retains a robust correlation with something we all desire: happiness. Countless studies have shown the same association each time. Those who feel intrinsically more connected with Nature generally enjoy greater life satisfaction and a more enriched sense of purpose.[278]

This chapter explores why this might be. And, more importantly, why it matters.

Feeling connected to Nature

Recent discussions with friends and family have emphasised how the word 'Nature' means different things to different people. Therefore, before I go on, it is incumbent on me to offer my own definition of Nature.

To me, Nature encompasses any lifeform that exists within an interacting network of beings, environments and biochemophysical processes. You may have noticed the inclusion of a human symbol in the Nature circle. By my definition, Nature contains us.

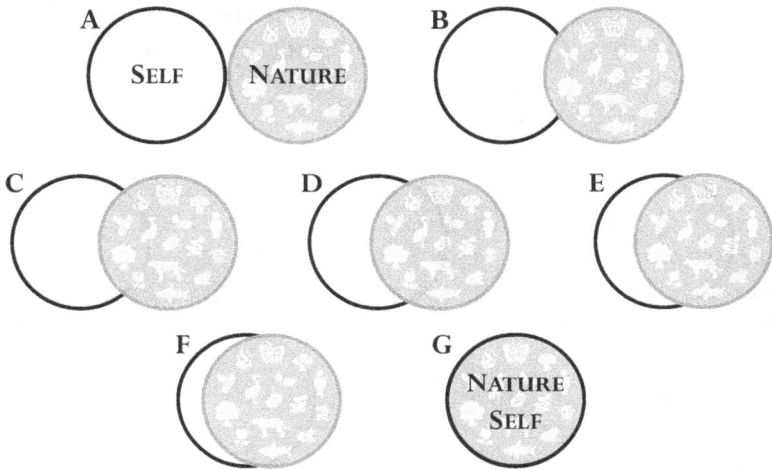

Figure 8. The Inclusion of Nature in Self scale: a visual scale of Nature connectedness from lowest (A) to highest (G).

More often, though, Nature is considered synonymous with wilderness. Implicitly, this serves to sever modern human civilisations from the natural world they are inextricably part of. Indeed, modern definitions of Nature refer to an entity that exists 'independently of people' or 'as opposed to humans or human creations'. This reinforces a 'take it or leave it' attitude when it comes to Nature. Vitally, the concept of Nature connectedness encourages us to reconsider this widely held stance.

To this end, more detailed measures of Nature connectedness have emerged over the years. Here, I highlight one that remains reasonably concise, known as the Nature Relatedness-6 scale. To work out your score, first consider how you relate to each of the following six statements:[279]

1. My ideal vacation spot would be a remote, wilderness area.
2. I always think about how my actions affect the environment.
3. My connection to nature and the environment is a part of my spirituality.
4. I take notice of wildlife wherever I am.
5. My relationship to nature is an important part of who I am.
6. I feel very connected to all living things and the Earth.

For each statement, give yourself a score from 1 (strongly disagree) to 5 (strongly agree). Then, divide your summed total by six to find your average response out of 5. Unsurprisingly, a higher score indicates a stronger personal connection with Nature.

When we examine this scale in more detail, we can see that the statements have been devised to address two core themes. The first is something I focused on at the end of Chapter 5, when I introduced the contrasting parenting styles of the Parent Bug and the Dark-edged Bee-fly: identity. Statements 3, 5 and 6 are the best examples of this. The second theme concerns one's experience, as emphasised in statements 1, 2 and 4.

But what I find most staggering is this. In revealing how you experience and identify with Nature as part of this brief scale, you unwittingly expose how fulfilled, autonomous and healthy you are likely to be. A more naturally minded life is consistently a happier life.[280]

The obvious question to me is why this should be the case. Is it that a greater baseline level of happiness induces a stronger connection with Nature? Or is it the other way around, in that a stronger connection with Nature brings about happiness? Or are they both inextricably tied up, like the seed and the tree, as one inseparable package?

Biophilia

E.O. Wilson was an American biologist and naturalist who hypothesised that strong evolutionary pressures would have favoured those of our ancestors with a greater affiliation to Nature. This is for the simple reason that our biological development took place in a pre-modern world, when the natural environment was intimately entwined with our daily survival.

When the ability to forage, hunt, protect and find shelter dictated not only your prospects of survival, but also, by extension, your reproductive success, you took much greater notice of your natural surroundings. For those who proved to be the most successful, a sense of contentment was evolution's internal reward. Wilson referred to this deep-rooted connection with Nature as biophilia.[281]

If the biophilia hypothesis is true, then it should be tightly wound into our DNA. In effect, there should be genetic signatures that distinguish those with high levels of Nature connectedness from those with low levels. While I am not aware of any study that has yet identified such a set of genes (or even looked for any), one study has instead investigated the heritability of Nature-oriented traits in a large cohort of twins.[282]

This study's method relies on the fact that identical twins share 100% of their DNA with each other, while non-identical twins share just 50%. Based on this, it is possible to estimate the heritable portion of a given behaviour. For example, let's assume that the desire to visit the lush rainforests and sublime naturescapes of Costa Rica before your 30th birthday is entirely decided by your genes. In this scenario, you would expect pairs of identical twins to be entirely concordant in their youthful, intrepid aspirations, whereas pairs of non-identical twins would only agree half of the time on average. Conversely, if this behaviour was in no way linked to your genes, then you would expect the same rate of agreement across both twin types. Notably, the validity of this method assumes each pair of twins is exposed to the same upbringing, thereby ensuring the effect of nurture is as equal as can be.[283]

What this study into the heritability of Nature-oriented traits found is that just under half of our innate drive to experience Nature is explained by our genes. The fact that there is any heritable element at all supports, at least in part, the biophilia hypothesis, but I acknowledge the theory is hardly home and dry... yet.[284]

The mechanism of Nature connectedness

It may at first seem strange to go hunting for genes that connect us to Nature. After all, this is an entirely different scenario to the case of spinal muscular atrophy, which I discussed in Chapter 3. In this severely life-limiting disease, scientists were compelled to go searching for the causative gene because the disease's impact on affected children was so devastating. The discovery of the *SMN1* gene in 1995 led to new knowledge about the disease's underlying cause, resulting in an effective treatment 21 years later.

Ultimately, this medicinal innovation hinged on a mechanism. In biology, we use this term to describe the natural processes that underpin a given biological phenomenon. For example, the mechanism underlying rickets is childhood Vitamin D deficiency, while the mechanism leading to Parkinson's disease is depletion of dopamine in movement initiation centres in the brain. In the case of spinal muscular atrophy, the mechanism involves the depletion of a protein that is integral to the survival of motor nerves. Mechanisms teach us how diseases tick.

But mechanisms are also vital in health. The mechanism underlying the Common Chameleon's colour-shifting skin is one based on guanine nanocrystals. The mechanism behind the Asian elm tree's successful defence against native parasites involves honed structural and biochemical adaptations. More widely, then, mechanisms teach us how the whole of biology ticks.

So if we are to fully understand the interconnection between healthier, happier humans and Nature, we must uncover its mechanism. This is where the identification of a genetic signature, if indeed it exists, would be invaluable. Analysis of the key genes involved would begin to teach us about the mechanism of Nature connectedness. Is it mediated through genes involved in the immune system, cellular metabolism, DNA housekeeping or sensory pathways? Or something else altogether? It is sure to extend beyond one pathway operating in isolation. The mind boggles to think of all the possible ways Nature has forged the unbreakable connections that unify it.

The fact that such in-depth genetic analysis is yet to be performed speaks more to the recency of Nature connectedness as a bona fide scientific construct than to the analysis' lack of scientific value. Instead, the potential mechanisms underlying Nature connectedness have so far been investigated in other ways.

First, though, it is crucial to point out an important subtlety that may seem somewhat counter-intuitive. Being 'connected' with Nature is not the same as being 'in contact' with Nature. You can't simply spend more time in green spaces and automatically expect to benefit from all the advantages that feeling more connected with Nature brings. One key reason for this, as I discussed in Chapter 6, is that many of our modern green spaces lack the biodiversity they used to boast. Instead, these spaces are geared around activities that not only distract us from whatever biodiversity remains, but they also disrupt the health of an already stretched ecosystem.

Don't get me wrong. Spending more time in green spaces *is* a beneficial thing to do. We know that it improves physical health, sustains mental well-being, boosts energy levels and lowers blood pressure. We just can't necessarily extrapolate those underlying mechanisms to explain why a closer *connection* with Nature is also beneficial.[285]

Despite this, exploring how our bodies respond when we are in contact with the natural world remains a very useful place to start. Up next, I look at some of the remarkable ways we sense Nature without even noticing.

Walking among the fractals

There is a black-and-white drawing by the British artist David Shrigley that tells us a lot about how we see the natural world. It depicts a tree that forms the same branching pattern at multiple scales. 'MAGNIFI-CATION REVEALS NATURE TO BE BORING,' the piece proclaims.

This conclusion is difficult to contest, when one appreciates that Nature, in many cases, is tediously repetitive. From the delicacy of a snowflake falling on Christmas Day to the shelter of a large oak tree, the apparent complexity of these contrasting forms is an illusion. Instead, our brains are easily fooled by the recurrence of simple motifs at varying depths of magnification.

Take a look at the snippet of winter woodland in Figure 9. As your eyes scan from the trunks to the twigs, absorbing successively more numerous divisions at shrinking scales, the more complex the image appears. There is even something slightly hypnotic about it. But in reality, I constructed these trees using just one simple pattern: the letter **Y**. Painstakingly, the trunks of progressively smaller **Y**s were layered over each of its predecessor's two branches to create a potentially endless set of divisions. If that explanation has left you feeling deflated, then you'll appreciate why the party magician rarely reveals his or her secrets.

What's more is that each tree in the background is formed by lopping off the first branch of the tree in front of it. In fact, each tree contains many self-similar mini-trees within. In creating this image, I have tried to convey the enticing charm of patterns that are independent to scale. These trees may be boring in one sense, but utterly mesmerising all the same.

This unique geometric organisation is an example of a fractal pattern. And it is rife in the natural world. While a snowflake and the woodland in Figure 9 are examples of *exact* fractals, the *random* fractals created by real tree branches, cloud outlines, coastal margins

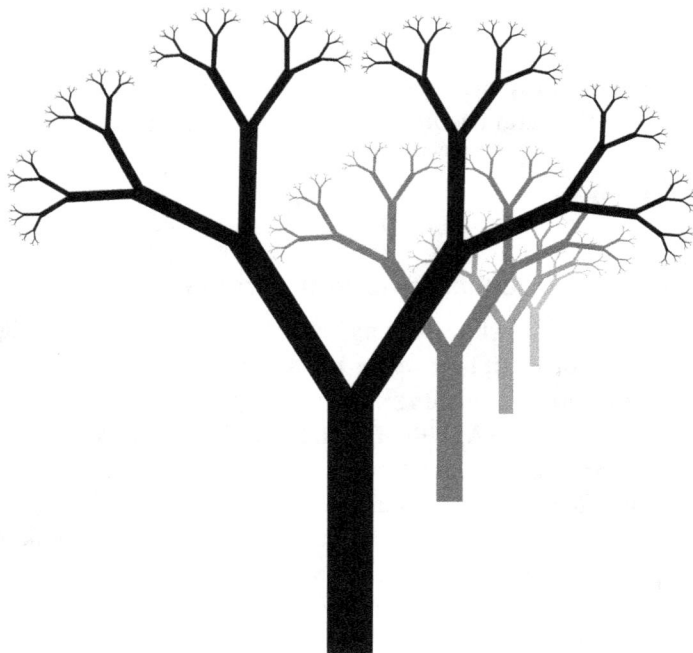

Figure 9. Woodland fractals: demonstration of Nature's predilection
for self-repeating, scale-invariant patterns.

and mountain contours are much more widely appreciated. However,
it is important not to misconstrue the 'random' descriptor as a less
rigorous process. As one zooms in and out, these natural fractals still
maintain their self-similarity, albeit in a statistical sense rather than
an absolute one.[286]

In contrast to the one-dimensional lines and two-dimensional
shapes we are so familiar with since pre-school years, a fractal is
described by a dimension that lies, rather unintuitively, somewhere
between 1 and 2. In essence, the fractal dimension measures the con-
volutions of a line as it breaches its smooth confines and begins to fill
the surrounding space. The closer to 2 it gets, the more complex it
appears.

And it turns out that the human visual system is rather particular
when it comes to fractals. Several studies have confirmed an intrinsic
preference for fractal dimensions between 1.3 and 1.5. This applies
not only to the aesthetic appeal of patterns, but also to the paths
our eyes take when viewing them. Most natural scenes and several

pieces of artwork, such as Jackson Pollock's splatter paintings, have been studied to show this effect. Furthermore, based on electrical recordings of the brain while individuals view a range of fractal patterns, it is these mid-range fractals between 1.3 and 1.5 that most effectively induce a state of restful alertness.[287]

Being simultaneously restful and alert may seem paradoxical at first. But given that the fractal dimensions of natural scenes cluster in the same 1.3–1.5 range, this frequently triggered brain state presumably served our ancestors well as they scanned their natural surroundings for food and foe. It is likely it provided the perfect balance between optimising energy expenditure and remaining vigilant in an often unpredictable world.

As a result of this evidence, an appealing idea presents itself. Among humans today, the visual system remains as perfectly in tune with Nature's fractal properties as it was for our restful, alert ancestors. Luckily, for those of us lacking the $£€¥ to purchase a fractal Pollock painting, a walk in the woods will provide the same calming boost for free.

How we learn to see

The fact that the human visual system resonates with Nature in this way tells just part of the story. Sadly, it has taken a modern pandemic for doctors and scientists to realise just how important our interplay with the natural world really is. And it is opening our eyes to a whole new dynamic.

During the third week of foetal development, grooves appear at the front of the developing human brain. These grooves furrow inwards to form two cavities, each one associated with its own pit, cup and stalk. As these structures continue to develop, two rudimentary eyes emerge. By ten weeks, each pit has formed a light-bending lens, each cup has created a light-sensing retina and each stalk has matured into an optic nerve, the ultimate conveyor of messages from eye to brain. In addition, many components, from ocular muscles to tear glands, develop alongside these structures to ensure full functioning of each eye down the line.[288]

But despite its early origin in foetal development, the human eye remains somewhat immature at birth. After all, the womb, devoid of light, is not exactly the ideal place to nurture such a light-sensing organ. At birth, the eye is only 60% of its eventual size, and its ability to resolve the world's intricate detail is diminished by as much as 60 times. Ultimately, the eye requires a wide range of visual stimuli in its early years to guide its maturity. In parallel, complex onward connections within the brain's visual processing networks are necessarily honed to provide efficient binocular perception of light.

This refinement process is reliant on a degree of focus. As light bounces off an object of interest and enters the eye, it subtends an angle dependent on the proximity of the object. While light waves bouncing off nearby objects are divergent as they enter the eye, those from faraway objects arrive in parallel. Despite this, the eye must find a way to focus near and far objects alike on the light-sensing retina. What's more is it must be able to switch between these tasks quickly, seamlessly and reversibly.

To understand how this is achieved, let's briefly consider the journey light takes as it travels through the eye. First, light traverses a transparent outer layer known as the cornea. Then, it passes through the pupil and navigates the solid lens, before finally shining on the retina at the back of the eye.

It so happens that as light crosses the corneal threshold, the air to cell interface alters its direction of travel. This is a fixed physical property of light that cannot be modified by the eye. But what the eye *can* modify, and indeed exquisitely control, is further deviation of the light waves as they pass through the lens. This is known as the accommodation reflex.[289]

Essentially, a fatter lens bends light to a greater degree, and this counteracts the divergent light waves arriving from a nearby object. This is associated with the eyes turning inwards and the pupils getting smaller, both useful when focusing on nearby objects. In contrast, a thinner lens bends light less and is perfect for the parallel light waves arriving from faraway objects. Cleverly, the accommodation reflex exists in healthy eyes to ensure this all happens flawlessly.

So the cornea and lens duo divert light by just the right angle to focus it on the most sensitive part of the retina. But to achieve the dynamism and accuracy it does, this corrective system is constantly detecting and acting on error signals. And given this system's objective is to produce a focused image on the retina, the error it responds to is blur. When a blurry image is detected by the visual processing areas of the brain, corrective signals are sent to muscles controlling lens shape, eye position and pupil size, thereby bringing the blurry image into focus.

Despite its ingenuity, the workings of such a dynamic system on a second-to-second basis are fairly relatable. However, there is a separate effect of blur in the modern world that is not only much less obvious, but actively harmful.

When our eyes are deprived

As I have already emphasised, the human eye is immature at birth and requires visual nurturing during childhood. The possession of a dynamic system that switches focus between nearby and faraway objects at will is pretty useful to modern humans, in many cases vital. And for our distant ancestors, this ability would have been a frequent matter of life and death.

But this is equally true for all vertebrates, even if the precise mechanisms depend on the animal's preferred environment. For example, when it comes to aquatic animals submerged in water, the air to cell interface at the cornea is absent. Therefore, in contrast to their terrestrial cousins, these animals rely almost entirely on the lens for their light-bending needs.[290]

Importantly, the majority of land animals (including humans) are calibrated to focus on faraway objects at rest, only activating muscles in control of lens shape, eye position and pupil size when viewing nearby objects. And it is *this* property of the human visual system that relates to the modern pandemic I hinted at earlier. But this pandemic is probably not the one you were expecting. This is the pandemic of nearsightedness.

In 2010, just under 2 billion people (28% of the global population) were nearsighted. Staggeringly, this is projected to amplify to 4.7 billion people (50% of the predicted global population) in 2050. This isn't just about who needs glasses or contact lenses when viewing distant objects. If inadequately treated, nearsightedness can lead to blindness in adults owing to its long-term effects on internal eye pressure, retinal integrity and lens transparency.[291]

It has therefore become of paramount importance to establish why nearsightedness is on the rise. The evidence consistently points to one factor in particular: children and teenagers are spending less time outdoors. And this is physically elongating their nurturing eyes.[292]

The mechanism behind this phenomenal observation involves the following. By spending more time indoors, children and teenagers focus on nearby objects more often. This causes more frequent detection of blur by the brain, which leads, as expected, to greater activation of the accommodation reflex. However, over time, each eye adapts to this persistent blur by physically stretching itself. It does this to reset the resting state of the eye, thereby reducing both the amount of blur and the required muscular workload when viewing nearby objects.[293]

Unfortunately, this effect is exacerbated by the fact that indoor artificial illumination is generally dimmer than outdoor natural sources during the daytime. What this means is that a nearby object viewed indoors typically induces more blur than if it were viewed outdoors.

Ultimately, the ocular elongation that occurs is irreversible. Without specialist intervention, it forces the parallel light waves from distant objects to be for ever focused *in front of* the backwards-shifted retina instead of *on* it. In essence, the unnatural lifestyles of today are permanently damaging our children's views of the natural world, both literally and figuratively.

This is not to say that every child who prefers the indoors will become nearsighted. Similarly, not every youthful lover of the outdoors will be spared this affliction. Like so many diseases, the environmental exposures imposed by our lifestyles do not act in isolation. Instead, they interact with our genes in ways that remain poorly understood.

Indeed, genes clearly play their role in this condition. For example, it is more likely for a child to be nearsighted if they have two parents with nearsightedness rather than just one. Implicated genes produce proteins involved in eye growth, chemical recycling in the retina and nerve signalling, among others.[294]

This illustrates a fundamental concept relevant to the study of human biology. While researchers often target the genetic underpinnings of a disease or its environmental associations in their scientific studies, it is most commonly a complex set of gene–environment interactions that matter most. And thinking about the main theme of this chapter, there is no reason to suspect that the mechanisms underlying Nature connectedness are any different.

Basking under the canopy

So far, I have focused on the visual aspects of our interactions with Nature. I have highlighted how our brains resonate with Nature's fractal patterns and how seeing less of the Great Outdoors is making a huge number of us *literally* see less. But I wanted to dig a bit deeper and discover whether there were wider knock-on effects elsewhere in our bodies. This led me, albeit only in a metaphysical sense, to Japan and the appealing local pastime of shinrin-yoku.

To stand among the trees is to embrace your senses. As my family and I scoured the coastal pine trees in southern Portugal, did it matter that we never managed to locate the Common Chameleon? And as I stepped tentatively closer to the wailing whistles I heard within a Quebecois woodland, would it have mattered if the Veery producing such notes had flown off before I set eyes on it?

For those advocating the benefits of shinrin-yoku, or forest bathing, these individual parts of the whole experience matter but a little on their own. It is complete immersion that counts. A twig cracks, a leaf falls, a bird cheeps, a breeze gusts, a beam glows, a log rots, a mist seeps, a vole darts, a stream flows...

A wealth of data now exists on the expansive health benefits of basking under the canopy. It improves cardiovascular health through reductions in blood pressure and heart rate, diminishes stress as evidenced by lower cortisol and adrenaline levels, and induces better sleep patterns. But there is one effect that I found all the more surprising.[295]

Shinrin-yoku boosts specific markers of immune function related to the functioning of so-called natural killer cells. These specialist cells are known to help combat viruses, bacteria and cancers. In a series of studies, a single weekend trip to forests of Japanese cedar, beech and oak trees was enough to elevate natural killer cell activity for up to a month. Crucially, this effect was absent in individuals spending a comparably active weekend in the nearest city.[296]

It is not certain how simply spending time amid the trees and their wild associates exerts such a positive effect on us. One prevailing idea is that natural aromatic oils, such as limonene and alpha-pinene, are released by certain trees and sensed by the human olfactory system. And so, as you walk under the canopy, your olfactory system is busy deciphering whichever of the trillion discernible combinations of stimuli are present, whether you consciously perceive them or not.[297]

In truth, I am sure that the mechanism is much more complex, assimilating subtle and often subconscious inputs from all our senses. In all likelihood, these then activate ancient physiological networks that only survived within our genes *because of* the enhanced survival they bestowed upon their hosts time and time again. These are the virtuous circles that whirl within our genes from one generation to the next, just waiting for the right environmental trigger in each of us to unleash their potential.

Natural shortcuts

At the age of 16, having just completed our first major set of school exams, a friend and I set off for Derbyshire, UK. And with a lightweight tent, several maps, a few days' worth of rations and a big dose of naivety, we felt ready to conquer the world. Little did we know that the route on which we had chosen to celebrate our new-found freedom was set to erode this balustrade of youthful enthusiasm down to little more than a grainy pile of grit.

Covering 268 miles, eleven counties and a total elevation gain rivalling that of Mount Everest, the Pennine Way, stretching between England's Peak District and the Scottish Borders, is a challenging and rugged walk along the island's backbone. The first and last days stick in my memory the best. I can still picture getting lost at the end of the first day, adding several miles to a route that had already taken us up to Kinder Scout, the highest point in the Peak District. I remember the final day, as we navigated the 25-mile slog across wild, featureless border country towards our first night's sleep in proper beds for a fortnight.

Why am I telling you this? The first reason is that it exemplifies an important feature of what makes us human. It highlights the fallibility of our memories. It turns out that we tend to remember the starts and ends of events more than what happens in between. Undoubtedly, my friend and I experienced a great deal more during the middle 13 days of our trip than we did on the first and last, yet most of what remains of that intervening period is, for me at least, a blur.

This is reminiscent of some nineteenth-century observations on human memory. One of the earliest examples was recorded by an American scientist, Francis Nipher, in 1878. He demonstrated that on recalling a chain of six random digits, the first and last digits were hardly ever forgotten, while the middle digits were misremembered up to a third of the time. This has become known as the serial-position effect.[298]

The mechanisms at play are varied and not yet fully understood. They involve the increased attention applied to the start and end of tasks and the saturating capacity of our short-term memories, as well as the benefits of a longer rehearsal period. Of course, I am acutely aware that with two and a half chapters of this book remaining you are unlikely to remember any of this by the end.[299]

Tihs is smliair to aohtner itretninesg obtaorsevin. When we read, it is the first and last letters of a word that hold the most value. Jumbling the middle letters does not impede us a great deal. But this extends beyond reading. More generally, the visual system excels at inferring detail when there are gaps in the scene. One important example is how the brain accounts for missing information from the natural blind spot. This occurs because the optic nerve, in the place it pierces the back of the eye as it heads towards the brain, is devoid of light receptors. Instead of perceiving an intrusive dark spot just to the side of our eyelines, our brains fill in the gap based on an educated guess.

These mechanisms serve a pragmatic value, saving us time and resources in an imperfect world. However, they are far from foolproof. For example, would our brains interpret 'brlaey' as 'barely', 'barley' or 'bleary'? Would it read 'ptlaes' as 'petals', 'pleats' or 'plates'? In these scenarios, we would certainly plead to the phrase's context for help. Moreover, when certain diseases raise the pressure inside the skull and, by extension, the optic nerve, the blind spot expands to such an extent that its guesswork severely impacts vision. Clearly, these shortcuts possess limits to their usefulness.

When I explore the route of the Pennine Way now, I can easily cite how it traverses at least twenty Sites of Special Scientific Interest, three national parks, two national nature reserves and an Area of Outstanding Natural Beauty. Yet none of this is catalogued in my memory of this trip, at least not in this format. Maybe I paid insufficient attention during those middle 13 days, but I would like to think that my brain *was* absorbing these sights. It was taking the necessary shortcuts to preserve as much pragmatic value from that natural environment as possible, assimilating those inputs along innate networks of restorative fascination and culminating in the immense satisfaction that I, like my friend, enjoyed on completing the challenge.

An imaginative exercise

The second reason I am telling you about this teenage excursion is that I was recently invited back to Derbyshire to participate in another journey. However, this was a journey much less to do with its physical rigour and much more about its cognitive demands and revolutionary potential. Crucially, it remains a journey without a clear beginning or end, just an enduring middle that we must continually find ways to absorb and assimilate.

It was on this daytrip back to Derbyshire in June 2023 when I contributed to the Silk Mill Vision of a Nature Connected Society. Led by Professor Miles Richardson's Nature Connectedness Research Group and attended by representatives from the Royal Society for the Protection of Birds (RSPB), Forest Bathing Institute, National Trust, Lawyers for Nature, Green Minds, Natural England and Moral Imaginations, the meeting aimed to source renewed hope for the future.[300]

The afternoon exercise was all about creativity. In groups, we were inspired to construct an imaginary society, one based on an innate connection with Nature. We discussed issues around societal values, infrastructure, energy, food, education, services, finance, healthcare and recreation. We talked about encouraging 'free-spirited childhood exploration of natural spaces', building a 'resilient sense of well-being' and guaranteeing 'freedom to future generations'. To finish, amid smiles and jokes, each group presented its vision alongside a five-point manifesto for a golden age to come.

This might all sound farcically aspirational, and in some ways it was. But there was also something hearteningly tangible about it. For all of us in that room, it unleashed the connective power of our imaginations. And I was left pondering the following question. How important could imagination be when it comes to the mechanism of Nature connectedness?

Given that I have already emphasised the vital role of imagination in human cognitive development, the short answer to this question is: probably critically so. But a longer, more satisfactory answer would perhaps judge whether there is a *specific* connection between our imaginative powers and Nature that can bring about positive change. After all, as history has proven many times, our imaginations are just as proficient at conjuring up harmful visions as they are beneficial ones. In an attempt to tackle this, I turned to a decade of work conducted by the Nature Connectedness Research Group.

Completing the virtuous circle

One of the most influential takeaways from this group's work is that there are five major pathways to Nature connectedness: senses, beauty, emotion, meaning and compassion. Unsurprisingly, the elements that I have already introduced throughout this chapter demonstrate considerable overlap with these five pathways.[301]

For example, as part of the Nature Relatedness-6 scale, I unveiled the two core themes of identity and experience. Clearly, how we intrinsically identify with Nature will affect the emotions (including compassion) we feel when we interact with the natural world, as well as the deeper meaning we attribute to such interactions. Undeniably, our senses are the portal through which we experience the beauty of Nature.[302]

In addition, I have focused on the positive influences of fractal patterns, outdoor childhood recreation and forest bathing. While these were primarily addressed in the context of passive exposure, it seems logical that their benefits are reinforced when there is active engagement.

This brings me back to the somewhat counter-intuitive point that I made earlier in this chapter, which stated that being 'connected' with Nature is not the same as being 'in contact' with Nature. What I purposefully omitted then was the critical role that engagement plays. It turns out that an essential tenet of Nature connectedness is that it correlates with how attentively one *engages* with the natural elements of a green space.

The type of engagement you choose can take various forms and degrees. It can be as simple as writing down and researching three things you see on a walk or describing three sounds you hear to a friend or family member. It can be taking an extra moment to stare up at the meandering clouds or criss-crossing branches and appreciating the fractal patterns on display for their intrinsically restorative quality. It can involve bedding a row of pollinator-friendly plants in your garden and watching the interactions they make with the local wildlife throughout the year. It can involve buying a light-box and placing it out at night to better appreciate the seasonal variations of invertebrate species in your area. In essence, you can be as *imaginative* as you like.

So if imagination facilitates engagement, and engagement enables us to better identify and experience the emotion, compassion and meaning evoked by the sensual beauty of Nature, all of which feeds back on itself and nourishes our imaginations once more, then doesn't that complete the virtuous circle? Well, no, not quite. But thankfully, the part that is missing has also been comprehensively studied by the Nature Connectedness Research Group. The part that is missing is *action*.

Let me take you briefly back to the UK's 2023 *State of Nature* report that I introduced in Chapter 6. In spite of the worrying trends that rightfully make the headlines, the report is also steeped with action. One large initiative is the 200-year Cairngorms Connect restoration project, which aims to revive populations of native trees, birds and moths. Another is the RSPB's Hope Farm venture in Cambridgeshire, which showcases the agri-environmental benefits of sustainable intensification on biodiversity and carbon emissions. More focused enterprises include the successful reintroduction of the Chequered Skipper butterfly *Carterocephalus palaemon* in Northamptonshire after 42 years of local extinction, as well as the Amphibian and Reptile Conservation's Natterjack Toad *Epidalea calamita* recovery programme. What all these schemes epitomise is purposeful and positive action for Nature.[303]

Studies have confirmed that individuals with greater Nature connectedness engage in greater pro-Nature behaviour. They volunteer more often in Nature-based projects, they donate more money to Nature-based charities and they make deeper personal sacrifices in the name of our Nature-based planet. This not only renders these individuals more fulfilled through an internal mechanism that our genes and environments have been conspiring in for millions of years, but this behaviour also circles back around, leading to enriched biodiversity, resilient ecosystems and happier, healthier humans the world over.[304]

Finally, then, the virtuous circle is complete. And this is how we can strive to fulfil our own symbiotic role in what some are calling the age of the Symbiocene.[305]

But it's important to emphasise that this existential journey didn't begin with the Paris Agreement in 2015, and it won't end with Net Zero by 2050. This is for the simple reason that this journey, for all pragmatic intents and purposes, is both beginningless and endingless. And faced with a voyage of such distance, woe befalls the civilisation that is entirely seduced by its nearsighted misgivings.

Nature as a therapy

As a clinical researcher focusing on improved ways to evaluate novel therapies in motor neuron disease, I am regularly re-assessing how to achieve maximum impact. Emerging treatments in this field aim to conserve whatever neuronal function remains, and consequently an emphasis is placed on making an earlier diagnosis. In essence, where there is a deterioration in health, the first goal is to halt that process and prevent further decline. The earlier you can do that, the better. Health conservation is a pre-requisite for recovery.

In Chapter 4, I suggested that our approach to the clinical assessment of patients can be equally applied to our suffering planet. Returning to this concept, it is clear that the idea of health conservation holds relevance when it comes to the deteriorating state of our natural world. And by exploring the manifold benefits of Nature connectedness in this chapter, I have conveyed how greater levels of pro-Nature engagement, imagination and action combine to form a powerful remedy, not only for us as individuals, but also for the planetary ecosystems we depend on.

The best thing about all this is that one doesn't need years of medical school to learn how to listen to, examine and investigate Nature. The human blueprint possesses the necessary biological tools not by some coincidence but because they have evolved to function perfectly in the natural world. Nature connectedness is in everyone's domain, and it is ready and raring to go.

Hippocrates recognised the benevolence of Nature 2,500 years ago, recommending certain herbal remedies and green exercise to his patients. As we can only speculate what effect this might have had on his patients back then, I wanted to explore how Nature is being deployed in modern therapies. Having spent over 15 years working in the UK's NHS, I can safely say that Nature-based treatments are not commonplace.

In the first instance, I was particularly interested in the role of Nature to treat disorders in my own specialty of neurology. I suspected there were very few studies addressing this, as I couldn't remember it ever featuring in our training days, hospital guidelines or patient leaflets. Maybe that is because Nature has been shown to be of no use for patients with neurological disorders, or, far more likely, it had neither been considered nor tested in this context.

To help decide which it was, some colleagues and I set about performing a systematic review of the medical literature. This is where you formulate a specific set of search criteria to probe extensive databases of clinical studies. We included any clinical trial assessing a range of Nature-based activities to treat the commonest neurological disorders. The most credible studies were the ones that recruited patients in an unbiased way, included a placebo group and evaluated a variety of outcome measures.

We identified 17 studies overall, the vast majority of which (94%) evaluated patients with dementia. The most common Nature-based activities were gardening sessions, petal arranging and farm visits. While 88% of studies reported an improvement in at least one outcome measure, there was huge variation in the outcome measures used across the studies. These measures assessed quality of life, well-being, cognitive ability, fatigue, depression and anxiety.[306]

The most striking, but admittedly unsurprising, result was the low number of studies identified. For many neurological disorders that we commonly manage as neurologists, such as migraine, Parkinson's disease and multiple sclerosis, there were no studies identified at all. If we compare this with the 24 studies exploring the therapeutic benefit of mindfulness in these three conditions up to 2023, then we can begin to grasp how neglected Nature has been.[307]

I am not suggesting that Nature-based activities represent a panacea. But given their low cost, high accessibility and positive association with a growing number of health outcomes, it is clear that we need to explore their utility in much more detail. Notably, chronic neurological disorders coincide with heightened stress, anxiety and depression in many patients, all of which negatively impact quality of life. It is through reducing these aspects of any long-term disease that Nature-based pursuits may hold their greatest value.

Nature-deficit disorder

It so happened that one study snuck into the first draft of our review that shouldn't have been there. Its subsequent exclusion was not because of its lack of interest (far from it), but because it didn't fit our original criteria. Instead of assessing Nature as a disease therapy, this study explored Nature's role in the underlying disease mechanism. More specifically, it explored whether the amount of Nature exposure throughout life is associated with the risk of developing dementia.[308]

It is worth digressing here for a moment to explain the ingenious way that Nature exposure can be estimated for any spot on the planet. Devised in the 1970s, this well-validated method depends on the way plants and trees shun the colour green. This may at first seem like an odd statement, given that healthy photosynthesising leaves are green. But they only appear that colour to us because that is the part of the visible light spectrum they reject. In fact, it is red light that the chlorophyll within leaves preferentially absorbs in order to power photosynthesis.

Red happens to have the longest wavelength of light that humans can see. And immediately beyond that visible limit is infrared. But while we cannot *see* infrared, we still sense it; we feel it as heat. When a human heats up, an orchestrated sequence of events, from sweating to diversion of blood to the skin surface, serves to offload infrared from our bodies. The problem is that plants, as they bask in red light from the sun, are also prone to heating up owing to infrared. So, just like the colour green, plants have evolved efficient ways of shunning infrared in order to avoid overheating. The fact that photosynthesising plants *absorb* red but *reflect* infrared is the basis of a measure called the Normalised Difference Vegetation Index (NDVI).[309]

What happens is this. Satellites orbiting Earth detect the amount of red and infrared emanating from every spot on the planet. Then a simple formula is used to calculate the NDVI for each location. This formula boils down to dividing the difference between their reflectance levels (Infrared – Red) by their sum (Infrared + Red), which produces an index between –1 and 1. If negative, it indicates a spot that is incapable of supporting terrestrial plant or tree life (e.g., a road or a sea). If zero, it implies there are no leaves in an area that could support vegetation. And the closer to 1 the index becomes, the more verdant the green space. Or from the leaf's perspective, the more vermillion the red space.

Going back to the study that we eventually rejected from our systematic review, we can now appreciate how the NDVI was used. Based in Florida, USA, a quarter of a million individuals over the age of 65 were identified from medical records. Both the dementia status and local NDVI for each individual were collated and compared. The study found that the odds of having the most common form of dementia, Alzheimer's disease, was reduced by up to a fifth for those living in the greenest areas. One cautious interpretation of this is that the mechanism underlying Alzheimer's disease may be mediated, at least in part, by a reduced exposure to Nature.[310]

And it turns out that dementia is not the only condition to show greater prevalence in areas of reduced greenness. Lung infections, heart attacks and mortality of any cause all show a similar link. In addition, at least two large studies from New Zealand and Denmark have suggested that the rising rate of attention deficit hyperactivity disorder in modern children is due to lower accessibility to Nature. More generally, the idea that children nowadays suffer from a lack of Nature exposure while growing up has been dubbed Nature-deficit disorder.[311]

Looking ahead, it seems to me that we are faced with two main approaches to this critical issue. One is about preventing what is yet to be, while the other attempts to cure what has already come to pass. The first one ultimately requires a systematic overhaul of what constitutes a healthy and happy human society, while the second one, acting in isolation, is like tiptoeing in front of an advancing juggernaut. But from a pragmatic standpoint, it is this second approach that might start getting us that little bit closer to where we need to be. If we are to prove beyond doubt the perils of a Nature-neglected world over the long-term, then do we not need to first reconfirm the value of a Nature-connected world in the short-term? From a health perspective, this amounts to pitting Nature connectedness against our modern ailments. By conserving the health of all Nature, won't we also conserve the part within it that relates specifically to humans?

Prescribing Nature

One person who set the ball rolling in this direction was not a doctor or biologist, but a geographer-turned-architect. In 1984, Dr Roger S Ulrich published a short report in the journal *Science* that has gone on to receive over 9,000 citations, an astonishing number for an article limited to just two A4 pages. Put simply, he discovered that recovery after routine gall bladder surgery was faster in a room with a natural view. By comparing the post-operative courses of 46 patients, he showed that the half of patients who recuperated in rooms with a tree view through the window required fewer doses of strong painkillers and were discharged from hospital earlier than the half of patients looking out onto a brick wall. Maybe it had something to do with the fractal patterns of those trees.[312]

Whatever the underlying mechanisms for this and similar observations since then may be, this concept endorsed a shake-up in hospital design. From ensuring more natural light on hospital wards to boosting access to green spaces for patients and staff, the calming caress of Mother Nature is there to soothe. I recently worked at St Thomas' Hospital in Central London, where a new Thames-side garden has been installed, right next to a main corridor that plays birdsong all day long. There should be much more of this in our hospitals.[313]

But as important as it is, the design of our hospitals and other healthcare settings can only take us so far. In Chapter 3, I emphasised that the healthcare setting triangulates the disease, the dis-eased and the easer. And fundamental to the interactions between them are personalised interventions, recommended by the easer in an attempt to rid the dis-eased of their dis-ease.

While we might instinctively consider medications, procedures and operations as the most influential of our interventions, this is not always the case. The lifestyle advice that healthcare professionals offer to their patients, covering aspects of diet, exercise, smoking, alcohol intake and sleep, can be just as impactful in the right context. Recently, as I discuss with patients how best to manage their neurological symptoms, I have added Nature to this list. But given the current absence of evidence for this approach in neurological disorders, am I justified in doing so?

One thing we can do to help answer this is to extrapolate evidence from other areas of medicine. In this regard, we can look to a series of studies that continue to explore the value of Nature Prescriptions.

The first time I learnt about this concept was in connection, once again, with Derbyshire's Peak District. 'Health professionals in Derbyshire have become the first in England to use RSPB Nature Prescriptions to boost health and wellbeing,' an article on the National Park's website began. It was dated 4 January 2023. It went on: 'A Nature Prescription is a free-to-use, non-medical approach based on accessible, self-led activities that people can do from home, on their own or with others.' Most important of all, Nature Prescriptions 'aim to create lasting connections with nature'.[314]

Was the RSPB breaking new ground here? It certainly was for the UK, having already run two successful pilot schemes in Scotland. While the first capitalised on the ruralism of the Shetland Isles, the second accentuated the Nature that thrives in and around the urbanism of Edinburgh. In this second pilot, just over 300 patients attending five GP practices were prescribed Nature. Just under two-thirds of patients in the trial suffered from anxiety and/or depression, although there were many other reasons for enrolment too, including insomnia, stress and headaches. The prescription itself comprised a calendar of Nature-based activities that prompted participants to engage, imagine and act.[315]

Championing the specific wilderness nearby, the Nature Prescription for Edinburgh prompted participants to 'smell the fragrance of yellow gorse blossoms' or to 'visit Edinburgh's Seaside and touch the sea'. But there were also general recommendations equally applicable to areas further afield. For example, participants were encouraged to 'look for kestrels hovering in the sky', 'find a patch of wildflowers and … imagine you are an insect exploring every flower' or 'listen out for five curious sounds in nature'. On top of these acts of engagement and imagination, the prescription also promoted positive action. One suggestion for July was to 'help look after the Water of Leith' by taking part in a 'river clean up'.

Three-quarters of participants felt that they benefited from their Nature Prescription, and 87% confirmed they would continue to strengthen their personal connection with Nature. And while this pilot was not intended to go into greater detail, I couldn't help but wonder what impact this intervention might have for patients attending neurology clinics. Alongside conventional therapies, I wondered whether Nature Prescriptions might provide additional benefit for patients with migraine, Parkinson's disease and multiple sclerosis in just the same way that mindfulness had.

Consequently, I reached out to RSPB's Sarah Walker, who is leading the English expansion of the Nature Prescription project. But before I spoke to her, I wanted to learn more about the history of this approach.

Known over the years by a variety of alternative terms, such as 'green social prescribing', 'Nature-based interventions' and 'ecotherapy', the present-day construct of the Nature Prescription can be traced at least as far back as 1998. In one early study in New Zealand involving almost 500 patients, written exercise advice from general practitioners in the form of green prescriptions reduced sedentary rates more than verbal advice alone. And between 2001 and 2022, there were 31 clinical trials that assessed the effect of at least one type of green prescription programme on psychological or physical well-being. The concept was gaining traction, albeit slowly.[316]

When I spoke to Sarah on a video call, I asked about the strengths of the RSPB Nature Prescription programme. With up-and-running projects in Derbyshire, Norfolk and Hertfordshire, plus one more being prepared specifically for stroke survivors in South Yorkshire, Sarah confirmed that the expansion was progressing well. She emphasised that Nature Prescriptions are most beneficial as part of an ongoing holistic dialogue between healthcare professionals and patients. Through these conversations, patients are encouraged to integrate Nature into their everyday lives. This is what makes the difference.

As I reflected on this conversation with Sarah, I was struck by the lurking tragedy at the heart of it. If I have confirmed anything in my own mind through writing this book, it is that humans possess an innate affinity for Nature. Why wouldn't we? We are part of Nature after all. Yet modern life has constructed monumental barriers, both physical and psychological, that encumber not only our access to the wider natural world, but even our in-built craving to seek it out. Apathy prevails, and our connection with Nature weakens further. We become so preoccupied with the stressful bustle of modern life that we can no longer see the smouldering wood for the burning trees. Our species isolates itself from the evolutionary networks that sustain it, and while we might convince ourselves that we are adapting, we are maladapting instead. In doing so, we neglect our basic psychological and physical need to connect with Nature. And our position within the Symbiocene is threatened.[317]

In that sense, is our collective disconnection from Nature not a form of ecological loneliness? This is just like the prevalent social isolation of today, with the heightened mortality it confers on its lonely sufferers, but amplified to the species level. Disturbingly, on foreseeing what this might mean for humanity's prospects in the long run, even the collective power of our imaginations might run short.[318]

Therefore, in the fourth and final part of this book, I turn to the future. And I consider how we might go about resetting our relationship with the natural world. First of all, I look at how the human brain is both susceptible and resilient to crisis.

PART IV

RESETTING OUR RELATIONSHIP WITH THE NATURAL WORLD

Giant Peacock

CHAPTER 8

Our Natural Aversion to Crisis

Solvitur ambulando

Marvin is hungry. He has been singing his duet since dawn, defending his freshly secured sector from newly arriving rivals. And now his energy reserves are in need of replenishment. He keeps low among the shrubs, where his tonal brown feathers blend into his immediate surroundings, and he scans the ground for something to break his overnight fast. It is mid-spring, and the early morning air remains refreshingly cool despite the inevitability of the afternoon warmth. He feels calmest and sharpest near the damp of the woodland floor, which percolates the morning dew.

But Marvin is also weary. Only the day before had he completed his relocation to this perfect woodland spot overlooking the adjacent river-dissected lowlands of Québec. Two months earlier, he had begun his 8,000 km journey from Brazil, flying mostly during the coolness of the night. Thankfully for him, he timed it just right, returning to his breeding grounds just as new life is emerging in earnest from the grips of the Northern Hemisphere winter. But there is little time for respite if he is to satisfy his strong reproductive instincts this season. After all, he did not fly such a distance with idle intentions.

Marvin spots movement in the leaf litter nearby. Crawling in and out of the partially decomposed organic matter, a caterpillar of the *Zanclognatha* clan is looking for a site to pupate and metamorphose. She makes use of the fallen leaves, helpfully shed six months earlier by the surrounding alders, oaks and elms on the proviso that their nutrients are dutifully recycled into the soil pending subsequent recapture. This particular caterpillar is dusky brown with stripes down her sides, thereby resembling the dead leaves she is so reliant on for sustenance and shelter. She continues her exploration of the woodland floor to find the perfect place to encase herself and begin her transformation into a moth, blissfully unaware of the predatory danger lurking behind.

'I wouldn't even bother,' comes a low voice from up high. 'Try as you might, you'll never catch that caterpillar.' Marvin looks up from his secure position in the thorny shrubs, perplexed. He tries to locate the voice's owner, but the early morning light is not intense enough to resolve the distant shadows above.

'I can tell you're confused,' continues the mysterious, resounding voice. 'Let me try to explain.' Despite fixing his eyes once again on the caterpillar for fear of losing her, Marvin remains curious and lends his ears to the unknown woodland orator.

'You see,' declares the voice, 'that caterpillar you espy so intently is sprightly as it moves further and further away from you. You may think you're sprightlier still as you hop across the woodland floor, but therein lies the problem.' The Veery takes one step closer to an opening in the bush, ready to pursue his prey once this untimely distraction has passed.

'In the time it takes you to arrive within predatory reach of the caterpillar's current position, she will have moved further forward, remaining frustratingly out of range.' Marvin rustles his feathers in preparation for his launch. But with his intrigue piqued, he stays to hear a little more.

'Even if you continue your chase from that new position,' the voice goes on, 'the same thing will happen. By the time you've caught up with where the caterpillar *was*, where she *is* will be beyond you.' Marvin ponders the apparent logical sense of the argument.

'I can assure you of one thing,' the announcement continues. 'If you embark on this doomed mission, you'll simply waste valuable energy on an impossible task. The reasoning, I'm sure you'll agree, is hard to dispute.'

As Marvin intently listened, the voice sounded increasingly familiar to him. Finally, it dawned on him who the owner of the voice was. It must be Daisy up to her usual tricks, although he had to admit that she had disguised her voice particularly well on this occasion. 'Time and time again,' Daisy concluded, 'as close as you may get to your nutritious breakfast, you'll never quite get close enough.'

Marvin, with his hunger outcompeting his weariness, wastes no further time on what he now recognises as a trick. Hesitating not a moment longer, he drops himself onto the woodland floor and bounces along the ground towards his long-awaited grub. When he gets close enough, he rummages through the leaves with his bill until he locates the tasty caterpillar, promptly consuming her with well-earned satisfaction.

'Solvitur ambulando!' Marvin cries back towards the speech's shady origins. He just makes out the silhouette of the trickster he suspected: a Brown-headed Cowbird *Molothrus ater*. Indeed, it *was* Daisy. On taking flight from her branch, she lets out a frustrated high-pitched rattle, sure sign that she's not best pleased. If only she could conjure a more robust way to dupe this astute Veery!

An infinite chase

This story is my adaptation of Zeno's famed paradox about Achilles and a tortoise (Figure 10). As a Greek philosopher living in the fifth century BCE, Zeno regularly played on the themes of motion and infinity. In one of his most widely known paradoxes, fast Achilles is told he will never overtake the slow tortoise despite his superior speed. This is for the simple reason that any chase he embarks upon can be divided into an infinite number of stages. And since an infinite chase can never end, the target remains permanently out of reach.

And yet, just as Marvin the Veery shrewdly proves, we all know that this is not true. To help resolve this paradox and give ourselves any chance of working out its relevance to this book's core message, let's revisit the fractal winter woodland from Chapter 7. At that time, I explained how I layered the trunks of progressively smaller **Y**s over each of its predecessor's two branches to create those fractal trees. In essence, this created a potentially infinite set of divisions. And given each trunk is twice the length of each of its branches, it is possible to calculate the total distance from the foot of the tree to the infinitely smallest twigs, *ab initio ad infinitum*.

Let's start with the most obvious part. That is, the main trunk of the tree has a length of 1 trunk. It follows that the first two branches of the tree each have a length of half a trunk, the next set of branches each have a length of a quarter of a trunk, and so on. This goes on for ever, with the denominator doubling and the value of the fraction

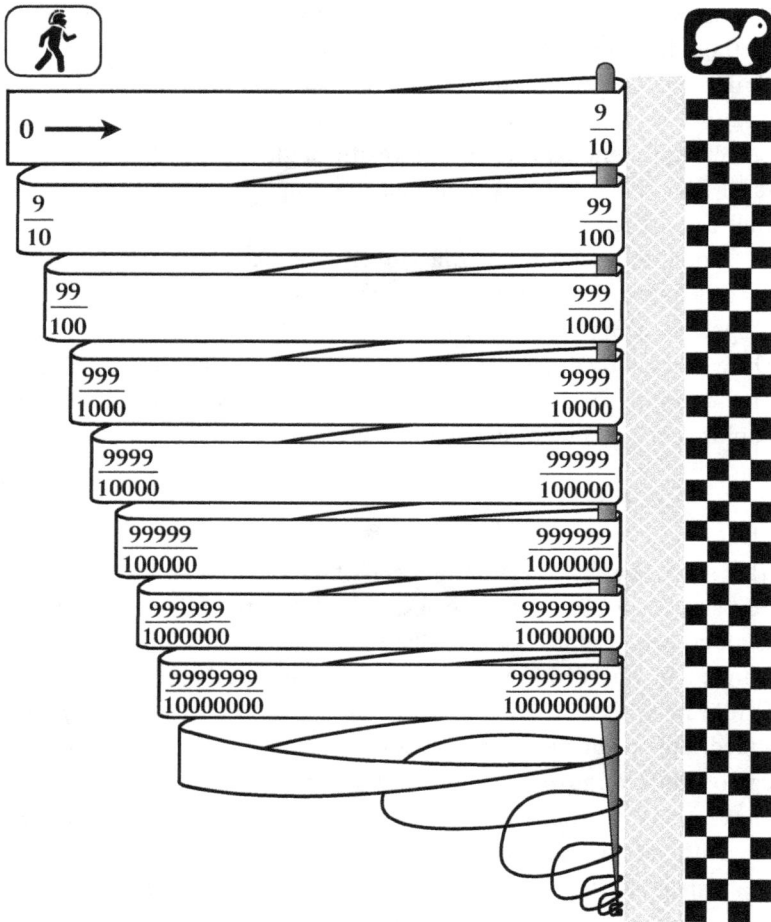

Figure 10. Zeno's paradox: As Achilles' chase can be split into an infinite number of stages, he can never overtake the tortoise.

halving each time. In summing this infinite series of lengths, you get $1 + \frac{1}{2} + \frac{1}{4} + \frac{1}{8} + \frac{1}{16} + \frac{1}{32} + \ldots$, which gets ever closer to, but never ever *exceeds*, two trunks.

You can think of it like this. Imagine you are an ant taking a journey from the bottom of the tree's trunk. By the time you get to the first branchpoint, you have covered a distance of one trunk. As an inquisitive ant, you wonder whether it is possible to travel the same distance again along the tree's progressively smaller branches. So you decide on the left branch and resume your journey. On approaching the second branchpoint, you feel you are progressing well, having reached the halfway point on your quest. This time, you decide to take the right branch and carry on. But as you get to the third branchpoint, despite being three-quarters of the way to your goal, you realise the problem. Whichever branchpoint you arrive at along the fractal tree, you can only ever add *half* of the remaining distance required to take you two trunk-lengths from the ground. In that way, it's impossible to ever reach your destination and you are eternally unsuccessful in your quest. What this illustrates is that the sum of an *infinite* process can indeed be confined within *finite* bounds.

This is why the Brown-headed Cowbird's declaration to the Veery was simply a ruse. It was a barefaced attempt at gaining a competitive edge over a neighbouring species through deceptive means. When Marvin proves the sum of such an infinite series is finite, he not only secures valuable calories, but also a psychological victory. Little does he know that the Brown-headed Cowbird more than achieves her retribution in the long run. In fact, it's worth digressing a moment to explain how this species pulls off one of the most devious tricks in Nature.

A female Brown-headed Cowbird makes no nest of her own. Instead, she lays her eggs in the nests of other bird species, including the Veery. In doing so, she is spared the resource-intensive task of rearing her own chicks, placing this immense ecological burden on her chicks' resourceful hosts instead. This so-called brood parasitism is well characterised in the avian world, perpetrated by the likes of cuckoos and cowbirds on many unsuspecting species, such as European magpies, certain warblers and sparrows, and, of course, Veeries.

But for now, at least, Marvin can revel in his recent triumph over the parasitising cowbird. Indeed, the valedictory flourish with which he vocalises his delight demands some further explanation. What exactly does 'solvitur ambulando' mean?

Something's afoot

In 1895, the English author, poet and mathematician Lewis Carroll, famous for his fantastical *Alice in Wonderland* and his nonsensical *Jabberwocky*, published *What the Tortoise said to Achilles*. When the tortoise sees that he has not only been overtaken by Achilles, but that his shell now serves as the warrior's resting place, he asks Achilles, 'So you've got to the end of our race-course, even though it *does* consist of an infinite series of distances?'[319]

Achilles replies: 'It *has* been done! *Solvitur ambulando!*'

It is solved by walking. This is what the Latin phrase *solvitur ambulando* means. This is what Achilles meant as he sat atop the defeated tortoise. This is what Marvin the Veery meant when he borrowed Achilles' ancient phrase deep in the Canadian springtime woodland. This is what Lewis Carroll felt best summarised the rational solution to Zeno's paradox of infinite motion. *It is solved by walking.*

But does this phrase not evoke much wider meaning besides? Does this phrase not speak to the benefits of the written exercise advice handed out by New Zealand doctors in the late 1990s, as well as the Nature Prescriptions currently being championed by the RSPB in the UK? Does this phrase not resonate in some way with E.O. Wilson's biophilia hypothesis, the stress reduction theory and the attention restoration theory all combined? Does this phrase not endorse the restorative benefits of shinrin-yoku in Japan, the sensory legacy of a Veery's birdsong in Canada and the enduring satisfaction of a 268-mile hike between England and Scotland? Does this phrase not embody the evolutionary fallout of our species' entwined cognitive and physical development, as our ancestors began to walk upright in the African savannahs several millions of years ago?

Maybe it does, maybe it doesn't. I'll let you decide how far to take it. But if one thing is certain, it is the following. This phrase *solvitur ambulando* symbolises a journey, one that for Marvin the Canadian Veery and Achilles the Greek hero ended in triumph, but for the devoured *Zanclognatha* caterpillar and the almost flattened tortoise ended in defeat. Writ large, these theoretically infinite journeys push their respective protagonists closer and closer to their natural endpoints. Consequently, it is to one of these endpoints, *crisis*, and the paths we take in arriving there (or not), that the rest of this chapter turns.

A health crisis looms

After that preamble, let me now dive in and illustrate what a human health crisis might look like as a result of a combined climate and biodiversity collapse.

There will be more medical emergencies caused by extreme temperatures, including heat stroke. There will be a higher occurrence of vector-borne diseases, such as malaria and dengue. There will be more crop failures leading to malnutrition and deficiencies of essential vitamins and micronutrients. There will be more physical trauma as a result of both turbulent weather events and military conflict. There will be a greater burden of mental ill health brought about by displacement, poverty and bereavement. There will be surges in respiratory illnesses owing to worsening air pollution. There will be heightened stress levels and low mood exacerbating the intensity of any long-term disorder. And there will be the emergence of new zoonotic pathogens as humans encroach on natural habitats without due respect.[320]

The grim upshot is this: such a health crisis will be plagued by a range of human diseases that will seem otherworldly.

But if we are to take anything away from Zeno's paradox, it is that there is always a different way to consider a dilemma. Rather than zooming in on crisis as a finite endpoint, should we not focus on the infinite steps that lead us there? In that sense, can we not *solve it by walking*? Essentially, this amounts to recognising our vulnerabilities *before* we hit crisis-point. The only problem is that this is rarely an easy task.

Showcasing a run-of-the-mill remedy

Let me exemplify this concept by highlighting the importance of an everyday vitamin. Riboflavin, also known as Vitamin B2, is a water-soluble vitamin ubiquitous in modern Western diets. Its highest concentrations are found in dairy products, meat, fish, green vegetables and grains. Riboflavin is essential for the normal functioning of several body systems, promoting the health of immune, skin, blood and nerve cells. It supports pathways that scavenge damaging molecules known as free radicals and stimulates mitochondrial energy production. From a practical standpoint, I regularly encourage my patients with migraines to trial riboflavin supplements, while the risk of stroke may be reduced in some individuals by boosting riboflavin intake.[321]

You can imagine, therefore, that when riboflavin intake dips below a critical threshold, disease can ensue. There is no greater proof of this than a rare neurological disease known as riboflavin transporter deficiency. Most commonly affecting children, its rarity in adulthood prompted my professor in 2015 to recommend publication of a case he had looked after in the preceding years. As a neurology novice at the time, I happily obliged.[322]

In just six weeks, our 35-year-old patient developed significant difficulties with her swallowing, speaking, breathing and walking. Notably, the quality of her diet had recently deteriorated as she coped with an abrupt separation from her husband. When she was first reviewed in the emergency department, she was at crisis-point and required urgent life-saving treatment.[323]

Thankfully, the medical team was able to support her breathing and nutritional needs, while the underlying neurological cause was investigated. Despite initially being subject to many inconclusive tests and several unsuccessful treatments, the clue to her diagnosis was a background of hearing loss over recent years. As is frequently the case in medicine, clinching the diagnosis came down to pattern recognition. Indeed, the pattern my professor recognised was a syndrome induced by insufficient riboflavin. Once this vitamin was administered to her at high doses, she began to improve. She recovered to such an extent that she was able to walk out of hospital without help a number of weeks later. I caught up with her in clinic several years on, and she was doing well thanks to daily riboflavin supplementation and the maintenance of a balanced diet.

The most remarkable thing about this case is that our patient's health crisis was remedied by a run-of-the-mill vitamin found naturally in a wide variety of foods. But why was she so susceptible in the first place?

The answer was found in her DNA. It turned out that she had mutations in both copies of a gene whose protein transports riboflavin across the gut wall and into the brain. With reduced activity of this transporter, even a short-lived drop in her dietary riboflavin intake over a few weeks meant her nerve cells bore the brunt. Those controlling her ability to swallow, speak, breathe and walk began to fail. Untreated, she would almost certainly have died. But with this simple treatment, her nerve cells were given the chance to recover.[324]

Despite this clear therapeutic success over a relatively short period of time, this series of events revealed something long-lasting. It disclosed a permanent vulnerability that was entirely unknown previously. Henceforth, the prospect of further health crises for our patient is only averted by her strict adherence to a riboflavin-rich regime.

A matter of life or death

From a conceptual standpoint, the theoretically infinite series of steps that culminated in this patient's definitive health crisis mirrors Zeno's paradox. But since the very nature of a paradox is prone to confuse, let me elaborate from a different angle.

Imagine 'crisis' as a cliff edge some distance ahead. Then imagine we are all blindfolded as we go about our busy lives, ignorant to the fact that cliff edges even exist. Since we are not aware of the precipitous danger ahead, we simply have no idea how close we are to it. And we certainly have no idea which step will finally take us over the edge.

The patient I've just described teetered on the edge of such a cliff, only to be relieved by swift medical intervention. This experience triangulated her location on the cliff edge and enforced a new-found wariness. But her underlying genetic risk of this particular health crisis had not changed, just her awareness of it.

In that sense, are we not all at some risk of crisis, blindly navigating some nebulous stage before we ourselves might teeter on the edge? If this is the case, then what we desperately need is an enhanced wariness of the danger ahead. But unlike our patient, it would be preferable if this new-found awareness were to arrive before we reach crisis-point itself.

To help develop this idea further, it would be useful to understand how the modern concept of crisis originated.

'Crisis' stems from the Greek word 'krisis', meaning 'decision'. Its origin in English dates back to the fourteenth century, when it was introduced, rather aptly, in a medical context. The earliest documented use of this word is found in Henry Daniel's *Liber Uricrisiarum*, which is an exposition of urine's role in managing the diseases of the time. Introduced as a period of great reckoning in the course of a patient's illness, crisis determined whether the outcome 'schal be gode or euel', ending either in 'lif or deþ'.[325]

Today, we still use the word 'crisis', alongside its related adjective 'critical', in certain medical contexts. An adrenal crisis, for example, is brought about by a potentially fatal inadequacy of the stimulating hormone cortisol in the bloodstream. Leading to dangerously low dips in both blood pressure and blood sugar levels, the failure of cortisol release by the adrenal glands can only be alleviated by prompt administration of a cortisol analogue. More generally, when a severely unwell patient's outlook is perilously uncertain, their condition is described as critical.

It is clear then that crisis encapsulates a set of key decisions that will decide between two opposing outcomes. In many cases, crisis decides between life and death itself. But why should we let it get that far if we can avoid it?

The conceptual shift I am proposing here is ultimately one of expectation. Our patient with riboflavin transporter deficiency had no expectation of her vulnerability to a specific and rare type of health crisis *before* it happened, therefore there was no way that she could have avoided it. But her critical experience in hospital changed that, and she can now make appropriate decisions in order to avoid further crises.

This approach incorporates reasonable expectations of the future based on real experiences of the past. It retains the original purpose of crisis as a time for making critically important decisions. But the crucial difference is that we now have a potentially infinite set of decisions we can make, all in the hope that crisis-point can be averted. I suppose we must thank Zeno for suggesting this approach might even be possible.

How we perceive a crisis

In the next section, I will look at some of the practical ways to avoid a health crisis caused by a combined climate and biodiversity collapse. But beforehand, I want to explain why I am reframing crisis in this way. At first glance, by suggesting we are on the path *towards* a crisis-point, yet to arrive at that definitive conclusion, I might even be accused of supporting the deferral of decisive action. In fact, this couldn't be further away from my intentions. As usual, I will rely on our species' position as an integrated part of Nature to outline the two main problems I am trying to tackle.

The first problem is that evocative words such as 'crisis' and 'emergency' are often extended to contexts that do not align with what our brains are telling us. Indeed, every living being exists in its immediate surroundings, and humans are no exception. Consequently, our brains have evolved immensely complex networks to assimilate and interpret inputs via our five senses. As I discussed in Chapter 1, this process dictates the emotions, thoughts and actions that serve as our brains' outputs.

The science is clear that wildfires, heatwaves, floods, storms and droughts are occurring more frequently owing to human-driven global warming. Yet unless you are in the midst of one of these extreme weather events, physically experiencing it for yourself, you will naturally distrust their labelling as a 'crisis' or an 'emergency'. Instinctively, this dissonance promotes internal conflict, and the identity you have carved for yourself throughout your whole life becomes threatened as a result.

Unsurprisingly, this problem is intensified in those who are distanced furthest from the immediate effects of climate and biodiversity breakdown. Based on the latest calculations, the wealthiest 1% contribute the same carbon emissions through their extravagant lifestyles as the poorest 66% do. Yet, living closest to crisis-point, it's the experiences of the poorest that resonate most strongly with the 'crisis' or 'emergency' label.

This leads onto the second problem, which is that evocative words such as 'crisis' and 'emergency' are too often perceived as an extreme viewpoint simply because of what they represent. It does not seem to matter that this viewpoint is based on a balanced scientific consensus that strengthens over time; its perceived extremeness serves to polarise. While I discussed this concept in detail in Chapter 5, there is one aspect I want to re-emphasise here.

The polarisation of viewpoints stymies the diversification of decisive action. I believe this is excluding a significant number of individuals who want to help but feel the only way to do so is through extreme personal sacrifice. They feel the only way to help the crisis is by completely giving up meat, planes and gas boilers in one fell swoop. They feel the only way to contribute to the fight is by climbing up a motorway gantry or slow-marching on a busy London road, thereby risking criminal proceedings. There is no doubt that thoughts like this excluded me for a long time, and I simply borrowed the most recent climate denialist story to protect myself from undesirable personal sacrifice.

But what I began to realise, as I increasingly appreciated our inherent connection with Earth's natural processes, was that the polarisation of viewpoints at this scale is wholly unnatural. Instead, it is the theoretically infinite number of decisions that our brains unconsciously make every second of every day that allow us to make a countless number of conscious decisions every day of every month of every year. Not all of these decisions need to be extreme. Not all of these decisions need to directly benefit planetary health. Some of these decisions can even retain the extravagances of yesteryear. But most of them should carry a reasonable level of expectation about our current path towards an ecological crisis-point. Most of them should be based on the intensifying wariness shared by climate and health experts worldwide. Critically, *all* of these decisions should celebrate the sensational diversity in all walks of life.

In doing all this, are we not satisfying our natural aversion to crisis?

But first comes a clean-up

As the Earth's climate and ecosystems continue to destabilise, leading all of us closer to a global health crisis, it seems obvious to me that healthcare personnel should be driving decisive action. After all, healthcare professionals wield a significant sphere of influence in society. If the health of our ecosystems is taken for granted by the vocational cohort that devotes its existence to sustaining human health, then isn't something amiss?

This is so important that health conservation should be part of every aspect of healthcare. I have already discussed one valuable approach, which endorses the value of Nature connectedness by prescribing Nature more readily. In Chapter 7, I highlighted how this simple device boosts pro-Nature behaviour through a mechanism of imagination, engagement and action, making individuals happier, healthier and more fulfilled at the same time.

But there's another vitally important way that healthcare can help. Leading by example, healthcare must become *less polluting*. Currently, if you rank global healthcare as a separate country, only four countries pollute the world to a greater extent. The challenge, though, is to reduce these damaging effects of healthcare while retaining the highest quality of both care and innovation. Is this too much to ask?[326]

In October 2020, NHS England released the first document of its kind anywhere in the world. In July 2022, its updated version was embedded into English law as part of the Health and Care Act 2022. Outlining a roadmap to the provision of net zero healthcare in England by 2040, the impetus for such a plan is clear. On the one hand, net zero will eliminate the 4% contribution that healthcare directly makes to England's total greenhouse gas emissions. On the other, it provides a clear message to other healthcare systems around the world, as well as to more polluting industries both at home and abroad: if *we* are adapting to protect global health, so can *you*.[327]

The range of strategies included in this roadmap is impressive. It covers medicine supply chains, emergency patient transportation and hospital heating, as well as the continued rollout of telemedicine. It underscores a plan to upgrade all hospitals to 100% LED lighting. It includes a commitment to optimise the environmental impact of single-use masks, gloves and aprons. It emphasises the importance of switching to anaesthetic agents and asthma inhalers that are more environmentally friendly, especially as these two groups of drugs contribute 5% of the NHS's total greenhouse gas emissions just by themselves.

But despite these specific aims, something else is made clear in the report. Given the high proportion of indirect greenhouse gas emissions that lie outside the direct control of any healthcare system, including how staff commute to work and how food is sourced for hospital canteens, the whole of society must become greener too. While healthcare will do everything it can, it can't do it by itself.

At the heart of this drive to clean up healthcare's trail of pollution is one of the cornerstones of modern medical ethics: non-maleficence (to do no harm). While its relevance in this context is self-explanatory, it is difficult to disentangle non-maleficence from the equally important ethical principles of beneficence (to do good), justice (to ensure fairness) and autonomy (to guarantee patient freedom). But for all the wonderful good that modern healthcare is capable of, this could all be quickly undone in the face of an uncontrollable health crisis. What's worse is this will happen unevenly on both global and regional scales, targeting those who are most vulnerable first. You can easily appreciate, therefore, that a health crisis brought about by climate and biodiversity breakdown weakens the very foundations of modern ethics. I, for one, find this incredibly unsettling.[328]

A missed opportunity

This brought me back to an idea I have had for some time. Since patients are rightly at the centre of everything healthcare strives to achieve, then don't we owe as great a commitment to the patients of *tomorrow* as we deliver to the patients of *today*?

Thankfully, when I started exploring this idea in more detail, I discovered I'm far from the only one to have thought this. Indeed, the UK Health Alliance on Climate Change, representing just over 1 million combined members of 46 health organisations, has proposed a brand-new component to the UK doctor's professional code of conduct. The theme of this proposal is sustainability, and its goal is to ensure that doctors practise 'sustainably to protect patients, the wider community, and the environment, both now and in the future'. This sounds like a no-brainer to me.[329]

Entitled *Good Medical Practice*, this professional code of conduct can be thought of as a modern-day Hippocratic Oath. It governs how doctors should behave at work and guides how doctors are appraised throughout their careers. It therefore casts a huge influence on the almost 400,000 registered doctors in the UK. And in preparation for its 2024 revision, the UK's governing body of doctors, the General Medical Council, undertook extensive consultation.[330]

Sadly, this sustainability proposal was not incorporated into *Good Medical Practice*, leaving many in the profession scratching their heads. Can we really afford to ignore this critically important facet of healthcare?[331]

But in reality this is not a widely discussed topic. This is partly because of a host of other pressures that health professionals have had to contend with in recent years. There's been the dire impact of the COVID-19 pandemic on patients, staff and resources across the world. There's been a global cost of living crisis. And amid an over-stretched and underfunded UK public health system, doctors, nurses and paramedics conducted multiple strikes in 2023.

Throughout all this, it has been difficult enough caring for *today's* patients, let alone *tomorrow's*.

Caring for the future

As I contemplated this more deeply, I became more curious about the potential experiences of tomorrow's patients. So I decided to consider two parallel scenarios, each one playing out in the year 2045. In the first scenario, sustainability sits firmly alongside the currently accepted tenets of doctorly professionalism: knowledge, skills, safety, communication and trust. In the second, sustainability finds itself continually excluded over the next two decades.

Scenario 1: Tansy hops off her bike and locks it inside one of the hospital's many cycle pods. As she proceeds down the tree-lined path towards her clinical department, a quarrel of house sparrows darts into the air. Strolling along, she can't help but admire the vibrancy of the freshly sprung daffodils beside her. Tansy's watch buzzes, confirming that her first patient has logged on to the hospital's newly upgraded telemedicine system. 'He's ten minutes early,' she thinks, as she arcs through the revolving doors while sipping the coffee she made at home. It may be early April, but she's grateful for the wave of cooler, ventilated air that welcomes her once inside. 'Good morning, Dr Marsh,' calls the receptionist. 'You're in your favourite corner room again today.' Dr Tansy Marsh smiles.

After entering the pleasantly bright clinic room assigned to her, Tansy brings up her list of patients on the computer. She opens the electronic records of the first newly referred patient to remind herself why a video call was most appropriate. 'Ah, yes,' she says to herself. 'Possible first seizure in a 29-year-old male. Neurological examination, blood tests and CT head scan were all normal in the Emergency Department.'

Tansy joins the video call and introduces herself as the neurology consultant. They discuss the events that occurred before, during and after the patient's recent collapse. Tansy confirms whether the details she has regarding the patient's past medical history, medications and tests are correct. 'I must admit,' the patient remarks, 'it's a wonder you have all the notes available to you. After all, this happened while I was visiting my unwell mum the other side of the country.' Tansy explains how tightly knit the electronic health system has become over the last decade or so. 'Thankfully, what this means is I don't need to repeat any of the tests already performed,' she confirms, 'although it's important we arrange a couple of other specialist tests to work out why this episode happened.'

Tansy outlines the two tests she recommends: a magnetic scan of the brain called an MRI and a brain wave trace called an EEG. Given both tests are frequently needed to fully evaluate the future risk of seizures, Tansy explains how they can nowadays be completed on the same day under the same roof. She highlights the patient's postcode on the electronic order form to ensure the nearest centre to his home is chosen. 'An electric minibus can even pick you up and drop you home on the day,' Tansy adds.

After emphasising important safety precautions that are relevant after a possible seizure, Tansy also reinforces the latest lifestyle advice from the Department of Health, entitled *Eat Well, Sleep Well, Walk Well, Naturally*. She goes on to explain that after just one episode like this, it's best to watch and wait without starting a new drug. 'However,' she emphasises, 'if you have another one, you must get in touch with me immediately to discuss treatment options.' The patient confirms his understanding on the screen. 'Let's arrange a video call follow-up in three months,' she says, 'but if the MRI and EEG haven't happened by then, we can postpone.'

'Finally,' she continues, 'if you're interested, I'll email you a leaflet about an active research study. We want to learn more about the interactions between sleep and seizures by measuring overnight body movements and brain waves in the homes of our patients. The modern devices are much less obtrusive than they used to be.' The patient agrees to find out more. And with that, Tansy ends the consultation and logs off. But before the arrival of her next patient, who requires face-to-face assessment, she takes a brief moment to enjoy a charm of goldfinches visiting the tree outside her window.

Scenario 2: Tansy is forced to circle the block three times before she finally finds a parking space in a side street. As she rushes between the slow-moving traffic towards her clinical department, two magpies quarrel over a half-eaten, discarded burger on the kerb. Dashing along, she can only hope her rushing doesn't provoke her asthma again. Tansy's watch buzzes, confirming that her first patient has already been waiting 20 minutes in the clinic waiting room. 'I'm late,' she curses, as she finally gallops through the permanently open doors. It may be early April, but she's grateful for the relief of the air conditioning once inside. 'Good morning, Dr Marsh,' calls the receptionist, 'I'm afraid you've been left with the closet room again.' Dr Tansy Marsh acknowledges the receptionist with a glance as she hastily fills a plastic cup at the water dispenser.

After entering the windowless room assigned to her, Tansy finds paper copies of each patient's referral lying on the desk. She scans the details of the first patient waiting for her. 'I see,' she mumbles to herself. 'Possible first seizure in a 29-year-old male. Neurological examination, blood tests and CT head scan were all apparently normal in the Emergency Department.'

Tansy invites the patient into the room and introduces herself as the neurology consultant. They discuss the events that occurred before, during and after the patient's recent collapse. Tansy asks about the patient's past medical history, medications and the tests that were done. 'I must admit,' the patient remarks, 'I thought the GP might have sent you my records. I know this all happened while I was the other side of the country, but they said they'd send everything through.' Tansy explains how the system is unfortunately not as connected as it could be. 'In reality, what this means is that I should ideally repeat some routine blood tests today,' Tansy confirms, 'as well as ordering some other specialist tests. That way, we can try to work out why this episode happened.'

Tansy outlines the two extra tests she recommends: a magnetic scan of the brain called an MRI and a brain wave trace called an EEG. Given these tests are performed by different departments, she explains that two separate appointments will be made, almost certainly on different days. 'As these are specialist tests,' she adds, 'they can only be performed in specialist centres like this one.' The patient says he understands, although he admits finding it difficult to get to the hospital considering he lives quite far away.

Tansy emphasises important safety precautions that are relevant after a possible seizure. She says it's safest if she prescribes an anti-seizure medication now, just in case another episode occurs. 'If you don't end up having another one, you can just dispose of it down the line.' The patient agrees. 'I'll arrange another follow-up here in three months,' Tansy says. 'Let's just hope the MRI and EEG have happened by then.'

'Finally,' she continues, 'if you're interested, here's a leaflet about an active research study. We want to learn more about the interactions between sleep and seizures by measuring overnight body movements and brain waves. We've set up a study in the hospital's research unit, where patients spend a week with us.' The patient agrees to find out more. And with that, Tansy brings the consultation to a close. As the patient leaves, the receptionist comes by to say the next patient is running 30 minutes late. 'Just my luck,' Tansy mutters.

Putting these two scenarios together helped to crystallise the myriad and often subtle ways healthcare can adapt to its sustainability challenge on a day-to-day basis. Not only can this be achieved without compromising patient care, but it is also an opportunity to revolutionise the experiences of tomorrow's patients. Two things are clear, though. Everyone needs to be on board with it. And it needs to pick up considerable pace.

Of course, medical science has, in another way, been impacting the outlook of future generations for some time. After all, the pursuit of scientific research is all about gaining new knowledge to shape the way we treat the patients of tomorrow. But my biggest concern is that much of this excellent research won't just be side-lined by the mounting burden imposed upon an increasingly fractious global health system; it will be completely obliterated by it.

Earth's least natural species

Before I move on to the final chapter, let me recap some of the key points from the second half of this book. Throughout, I have tried to dissect the deep-rooted position that our species, just like all others, occupies in Nature.

I first tackled biodiversity through an evaluation of the bacterial variety resident in our guts. I highlighted how the microorganisms present from the earliest stage of animal evolution have helped to shape the body plans of all animals that have ever lived. And by creating a bodily axis, the alimentary canal from mouth to anus has endowed its hosts with a fundamental sense of the differences between left and right, front and back, and head and bottom. Bilaterians, such as the primitive ragworm of today's Eurasian coastal waters as well as Earth's only surviving human species, have taken advantage of these anatomical gifts to diversify in response to their particular surroundings. All along these journeys, every animal remains reliant on its microscopic navigators, suffering immensely from their loss. In the case of humans, we only need to consider the neurological features of Parkinson's disease to grasp how interconnected our microbial guts and brains really are.

Bursting from my own fascination with moths and butterflies, in all their glorious diversity, a trip to London's Natural History Museum inspired me to re-explore the origin of species. I considered how a single ring species, namely the Ensatina salamander living at the fringes of the Californian Central Valley, is essentially a slow-motion film of two species in the making. For Linnaeus and Darwin, it was ultimately all about celebrating the ecological variety they observed around them that defined their legacies.

Next, I dipped my toes further into the swirling pool of biodiversity by looking at how it can be measured, modelled and manipulated, both in computer simulations and in the real-world grassland Jena Experiment. Crucially, I learnt that it's the richness and evenness of an ecosystem that dictate its resilience to environmental pressures and, therefore, its ability to evolve over time. But I was also left mourning the extinctions of those undiscovered species, whose fossils will be the only evidence to future scientists of their curtailed existence on Earth. In that vein, I summarised the parlous state in which our biodiverse world finds itself. And in catching ourselves red-handed, we must turn to our own green-mindedness to wash away the blood. Ultimately, the universal language of Nature depends on this change of heart for its own survival.

To spur this on, researchers have demonstrated the value of Nature connectedness as a means to boost human health and happiness. If the biophilia hypothesis is true, then our innate affiliation with the natural world not only acts as an indicator of our ancestors' ecological success, but it also remains as responsive to its underlying gene–environment interactions as ever. At one level, the mechanism for this involves our creative engagement with Nature, promoting positive pro-Nature action, although the physiological pathways at play remain somewhat elusive.

Despite this, we can look at how our bodies respond to natural stimuli in order to focus our minds. Whether it's how the physical structure of our eyes is nurtured during childhood, how our brain waves react to the most familiar fractal patterns or even how our immune systems are beneficially influenced by time spent in the woods, you simply can't lose sight of the fact that humans and Nature are one. And any deviation is prone to inflict an alienating loneliness on us all.

Then, in this chapter, I've considered lessons from Zeno and Carroll. In essence, they both teach us that life is a journey. But it really doesn't matter how finely you choose to divide and categorise that journey; Nature has already provided us with all the faculties we need in order to enjoy, understand and embrace it. Maybe the greatest challenges in life really are solved by walking.

Thus, we hit upon the greatest challenge of them all: how can one know what the future holds? In among all the roadmaps, cliff edges and futuristic visions remains a candid reflection that, of course, nobody knows for sure. Isn't this the greatest source of ambiguity in life? Isn't this what breeds the most conflict?

Nevertheless, the decisions we make on a personal, national or intergovernmental level must be decided somehow. So it makes sense to me to base these decisions on something pervasive, consistent and relatable. It makes sense to me to protect the tried and tested methods we have based these decisions on for hundreds of millennia. It makes sense to me to let our moral compasses and our powers of observation guide the way. Aren't our modern ethical principles and our modern scientific processes just the latest versions of these intrinsically natural qualities?

Finally, when it comes to our impending health crisis, I have proposed a shift in the way we approach this existential issue. Why do we feel disconnected from what the climate scientists have been increasingly sure about since the 1970s? Why is our collective behaviour as a species not as responsive as it needs to be? I believe the answer to these questions is, ironically, the same as what created the problem in the first place. Steadily, during our rise to dominance on this planet, we have excommunicated ourselves from the very thing that gave us the tools to fulfil such a task: Nature. There is no way this was intentional in the early days. Even now, I don't think this self-destructive streak is as purposeful as our recklessness might suggest. I think we have simply got caught up in a spiralling mess of self-perpetuating feedback loops that we are struggling to extricate ourselves from.

Sadly, we have become Earth's *least* natural species.

Putting a new spin on it

So what can we do about it? In proposing an answer, I return to the human brain. And I come back to the fact that our brains are exquisitely sensitive to their immediate surroundings. When mismatches occur between the labelled climate crisis and what our brains are telling us, it creates an internal conflict that cannot be easily overridden. Since that mismatch tends to be greatest in the temperate silos of our civilisation's rich and powerful, it gets internally labelled as extreme, and the world polarises further. For most of us, this global crisis always seems too remote and too extreme to get overly fussed about.

But there is one other aspect that is arguably even more important. As I highlighted in Chapter 1, our brains' natural sense of time is in sync with the cyclical rhythms on Earth, whether it be a day, a month or a year. We are in tune with the sunrise and sunset each day. We obsess over the seasonal variations of our localities. But as a consequence our brains do not process longer passages of time well. How often are you amazed that a certain event in your life happened much longer ago than you remember? And when you bump into a long-lost friend in the street, how difficult can it be to pinpoint when you last saw them? Indeed, these tasks fall outside our innate resonance with the natural cycles of time. I can only presume that a more accurate appreciation of linear timescales over many years would not have altered our survival a great deal. Otherwise, we would be much better at it.

Since this effect is evident over timescales much shorter than our own lifespans, imagine how infinitely magnified it becomes when we turn to the vast geological timespans that govern the workings of our planet. Our brains simply have no biological frame of reference for a thousand years, let alone a *thousand* thousand years.

This natural lack of appreciation for linear timescales is thwarting our appreciation of the crisis we are facing. This is for one simple reason. Without an instinctive awareness of longer passages of time, we have no instinctive awareness of the rate at which things are changing.

In Chapter 5, I highlighted how the speed of the Indian tectonic plate had been miscalculated owing to an error in the estimation of a particular time period around 67 million years ago. Well, it turns out that our brains are prone to the same type of error. This is not because they are at fault, but because they never learnt how to gauge the vast timescales involved. Indeed, they were never taught.

When we hear that the global average temperature has risen by 1.2 degrees Celsius in 150 years, we automatically think it doesn't sound like much. After all, we experience much larger swings in temperature on a daily or seasonal basis. We simply have no innate frame of reference for how quickly global average temperatures *should* be rising over that period of time.[332]

But at least when it comes to sensing air temperatures, we do have the requisite thermoreceptors in our skin. The problem is compounded when applied to changes in atmospheric carbon dioxide levels, the major driver of global temperatures. This is because our biological frame of reference for sensing carbon dioxide only operates at concentrations 100 times greater than atmospheric levels. Plus, it's internal blood levels we are able to detect, not external atmospheric ones.

We know that prior to the Industrial Revolution there were 280 parts per million of carbon dioxide in the Earth's atmosphere. But as I write this in December 2023, the value is 422. That's approximately a 50% increase in 250 years. Prior to this, the fastest that atmospheric carbon dioxide concentrations had risen by 50% occurred about 55 million years ago. At that time, it took about 1 million years to do it. This means that when it comes to such large jumps in atmospheric carbon dioxide levels, the rate of change over the last 250 years has been *4,000 times faster* than during any other spell since the extinction of the non-avian dinosaurs.[333]

Let me put this into a more relatable context. If the Earth span 4,000 times faster than we are used to, a day would last just 22 seconds! Now imagine if human activity were the cause of such a drastic change. Wouldn't this have spurred much greater action from us much sooner?

Put simply, our brains are poor natural barometers of global climate change. But why would we expect them to be any good in the first place? They have never had to deal with such a problem before. As a result, we will always miscompute the situation if we rely purely on our innate frames of reference. And the crisis will always seem too remote, too extreme and too slow.

Therefore, given our health crisis is playing out over distances and timescales way beyond our sensory limits, we need to recalibrate our instincts to bring us back in touch. Thankfully, we already possess the cognitive abilities to do just this. We can do it using the two major streams of information flow that have evolved wonderfully intricate ways of converting brain inputs into brain outputs: creativity and language. We can do it through an honest analysis of our delicate relationship with knowledge and disease. We can do it by re-engaging with the natural world that we are intimately part of. And we can do it by redefining 'crisis' as the pathway of infinite decisions we must now navigate in order to avert us from crisis-point.

And so, in Chapter 9, I finish by exploring what a reset for the natural world might entail.

CHAPTER 9

Homo insapiens

The power of sleep rests in awakening the mind

I f the Earth's changing climate teaches us anything, it is that a less stable, less biodiverse world represents a big test for humanity. And when facing up to this big test, there are naturally many visions of a big fix.

For those who agonise over the unabating conversion of Earth's subterranean organic deposits into at least 35 billion tons of airborne carbon dioxide each year, the big fix is about slowing down and growing down. For those who recognise ecological isolation as the foremost source of biological desolation, the big fix is about renaturalising our senses. For those who are critically aware of our overshot planet, the big fix is about renewing the regenerative capacity of our energy flow.[334]

But, living in the biodiverse world that we do, we should expect and embrace a range of alternative visions besides. Recently, I stumbled across an excellent resource produced by French graphic artist Léonard Chemineau and publicised by the Islington Climate Centre. Based on a 2020 article published in the journal *Global Sustainability*, it captively illustrates the 'discourses of climate delay'.[335]

In the main graphic, 12 schools of thought are summarised, incorporating groups such as the doomers, the talkers and the free riders. It separates the robotic faith of technological optimism from the deflective blame of whataboutism. It distinguishes the paralysing inaction of policy perfectionism from the everyday virtue of individualism.

But among these catchy terms, each form of delay boils down to the same core issue. In its own way, each one sparks the embers of ambiguity that enkindle conflict. As a result, the critical decision-making processes become warped, further fuelling the climate crisis.

This draws parallels with the ethical principles of beneficence (to do good) and non-maleficence (to do no harm) that I touched upon in Chapter 8. While these two principles conventionally relate to the ethics of human healthcare at the individual level, it is perfectly feasible to apply them at the population level too. Essentially, each discourse of climate delay not only has the potential to cause direct harm, such as the toxic persistence of fossil fuel solutionism, but it can also thwart the good otherwise implementable through non-delay.

In many cases, the false sense of security generated by these discourses further impairs our naturally poor ability to sense climate change. Ultimately, this false sense of security pushes our views of the climate and biodiversity crises to points in our minds that seem even more remote, even more extreme and even slower than they would have been otherwise.

So what's the solution? Well, maybe there is no single 'big fix' to reset our relationship with the natural world. Perhaps, instead, we should turn our attention to the cumulative power of many 'small resets'. After all, isn't this how biology often sorts out its problems?

Let me now share with you an example that I hope is as persuasive as it is pervasive.

Getting our heads around sleep

Like eyelashes after a tiring day, pairs of pinnules take twilight as their cue to reconvene a nightly union. Packed full of sugars thanks to the day's ruddy glow, the predictable cyclicality of Earth's spin ensures timely resumption of this leafy pact of rest and refuge. This is a biological adaptation that guarantees the cooler, darker hours out of sunlight's direct reach are spent in the successful resistance of evaporative and predatory threats. At the end of the day, this is an adaptation that allows certain plants, such as the Sleepy Plant, to close their small feather-like leaves and 'sleep'.[336]

There is no doubt that oddity invokes language through the medium of name. Therefore, when we consider *Mimosa pudica* not simply as the Sleepy Plant, but also as the Touch-me-not, Live-and-die, Shameplant and False Death, to list just a few of the names used across the globe, we can appreciate just how *odd* humans have found this evolutionary stalwart of the pea family.

But there is actually nothing odd or strange about the light- and touch-sensitive movements of the Sleepy Plant. As a native of equatorial Brazil, it exhibits the ideal behaviour for a plant that's both doused in light and housed in darkness in even measures. We only need to consider an alternative geological scenario to appreciate how ideal its behaviour is in a world such as ours.

Imagine for a moment that the Earth, as its spherical form coalesced from clusters of space dust and swirls of gas 4.6 billion years ago, had not acquired the seemingly perpetual momentum of a spinning top. Instead, imagine that the only movement it retained was its annual orbit around the Sun. What might a spineless planet have meant for the evolutionary prospects of the Sleepy Plant?

With transitions between light and dark occurring every 6 months instead of every 12 hours, the concept of a 'day' would not have existed. Instead, it would have plunged any equatorial life into the most extreme summers and winters. But what is unknown is whether such lengthy periods in constant sunlight or in unremitting darkness, with the incomprehensibly large swings in temperature and light levels this would have generated, could have permitted Earthly life to commence at all, let alone allowed anything resembling our notions of sleep to evolve.

My point is that Earth's 24-hour spin cycle, which means most places on the globe (save those nearest the polar regions) transition between light and dark 730 times every year, is firmly entrenched in all walks of life. But while you may not have ever considered whether plants sleep or not, we are all suitably acquainted with our own health's requirement for regular slumber. In fact, sleep is our brain's way of getting all the small resets it needs.

Anyone reaching their 80th birthday has probably spent at least 20 years of their life asleep. For some, it might even approach 30 years. But since sleep renders its host temporarily unconscious, this chunk of our lives vanishes into a dream-studded vault of nothingness. What we easily forget is that sleep serves many vital biological functions.

This should come as little surprise, considering the wide range of animals exhibiting some form of sleep. Birds, fish, reptiles and amphibians all sleep. Even insects, arachnids, molluscs, worms and brainless jellyfish all sleep. Like any behaviour that demonstrates such conservation across evolutionary classes, the very first sleeping animals must have enjoyed an immense survival advantage over their sleepless peers. These early animals were the pioneers of repose.[337]

However, it is important we do not view sleep as a time of passive inactivity. The truth is that sleep has evolved to be a highly regulated, purposeful and indispensable component of animal physiology. Importantly, it is a time for the consolidation of memory, conservation of energy, repair of DNA, removal of cerebral waste products and regulation of temperature.[338]

Humans exhibit two distinct sleep states, distinguished by the fact that rapid eye movements (REM) are observed in one, while they are absent in the other (non-REM). Although non-REM sleep produces electrical brain waves that are slower than those found in wakefulness, the brain waves occurring in REM sleep paradoxically resemble the awake state. From the moment we drift off, we cycle through alternating periods of non-REM sleep (encompassing light, medium and deep stages) and REM sleep. With each cycle lasting roughly 100 minutes, adult brains typically need four to five cycles for a restorative night's sleep. While the regions of our brains most active during the previous day engage in the slowest, most revitalising brain waves during non-REM sleep, it is the temporary fast waves of REM sleep that allow us to dream.[339]

And just like any bodily function, the value of sleep is best appreciated when it goes wrong. Given its complexity, it is no wonder that sleep disorders are the ultimate disruptors. Some people fall asleep at unwelcome moments, while sleep fails them at desirable times. In some individuals the normal limb paralysis of REM sleep is reversed, causing them to act out their dreams. But in others an unpleasant awareness of this normal limb paralysis temporarily locks them within their nightmares. Then there are all the sleeptalking, sleepwalking, head-exploding, bedwetting, night-terrorising, limb-moving, tongue-biting and breath-stopping descriptions that are frequently reported in both general neurology and specialist sleep clinics.

Perhaps most surprisingly of all, simply sleeping less or more than the Goldilocks window of seven to nine hours per day may confer a 12–30% increased risk of mortality. While John Steinbeck referred to sleep as 'death's brother' in The Grapes of Wrath, Mahatma Gandhi drew a similar comparison when he said, 'Each night, when I go to sleep, I die. And the next morning, when I wake up, I am reborn.'[340]

Despite the apparent vulnerability it bestows on its host, sleep shows itself to be a tireless adaptor when it comes to the art of survival. For example, some birds sleep during flight and some land mammals sleep while maintaining an upright stance. For such animals, sleep offers no barrier to remaining on the move or being ready to scarper from a predator at a moment's notice. Remarkably, certain birds and marine mammals can independently rest each side of their brain, literally allowing them to sleep with one eye open when surrounding dangers make this an advantageous tactic.

Such variety across the animal kingdom is evidence of one thing: each species has carved out its own survival-promoting way of regularly resetting its nervous networks. The fact that this is achieved through close synchronisation with Earth's axial spin is yet another example of how biology is constantly moulded by its environment.

Early birds and night owls

On closer inspection of our own species, we find considerable variety in sleep behaviour. There are the early birds who are first to rise. And there are the night owls who are last to bed. There are also a smaller group of more erratic sleepers whose internal timekeepers fail to adhere to a consistent cycle. Then there are those whose jobs and lifestyles mean they must periodically overcome the somnolent trials of regular shift work or jetlag. What this all points to is an internal body clock that exists in all of us.

Deriving from the Latin words 'circa', meaning 'about', and 'dies', meaning 'day', we all have a circadian body clock that synchronises our daily activities with the spin of the Earth. Set by a group of genes that maintains a near 24-hour cycle in the absence of environmental cues, the clock works best when reset each day by the presence of the most efficient *zeitgeber* (a German word, meaning 'time-giver'): sunlight.[341]

This topic reminds me of a 2019 trip I took from London, UK, to Perth, Australia, to present my main PhD results at an international conference. I had never travelled that far before, and so I thought nothing of giving my presentation on the second day of the trip. But after traversing the 9,000 miles and 8 time zones between London and Perth, I came up against the doggedness of my natural circadian rhythm. While still wide awake at 5 a.m. in a desperate spiral of sleep deprivation, as I reflected on how this was hardly the best preparation for my forthcoming presentation, I resolutely penned my unsettled state of mind:

> Time – itself – is nothing
> but a timeless
> formless structure
> where all human madness hangs –

While such unnaturally big jumps in sleep–wake patterns are clearly troublesome to modern sleepers (and their poetic abilities alike), a recent study suggests it is the small jumps in our evolutionary past that shed the most light on the fascinating variety of human sleep.

But before I comment on the main result of this study, I need to recap one important aspect of human evolution. In Chapter 4, I outlined how our species took part in an African emigration that began about 60,000 years ago. Unlike today, this was a time when *Homo sapiens* shared the planet with several other human species. Two of our phylogenetic cousins, the Neanderthals *Homo neanderthalensis* and the Denisovans (whose scientific name remains unconfirmed), had already been calling Eurasia home for several hundreds of thousands of years before our arrival there. Making a much earlier exit from Africa, these species had accustomed themselves to a major environmental challenge associated with life away from the equator: seasonal swings in the daily duration of sunlight.

Exemplifying the pervasiveness of gene–environment interactions, this turned out to be rather useful for our own ancestors arriving for the first time in Europe and Asia. Recent analysis reveals how interbreeding with these archaic human species led to the selected transfer of gene variants critical for resetting our circadian body clocks to the new Eurasian environment. And it's the early birds alive today, who favour an early night and a morning sunrise, that carry more of these Neanderthal and Denisovan gene variants.[342]

The implications of this are a real eye-opener. Despite their subsequent extinction, the genetic legacy of these two archaic human species lives on in the small resets that the body clocks of modern humans are making every sunrise and sunset, particularly in those living furthest from the equator. But consider this as well. Given it is the night owls alive today, with a default preference for later bedtimes and longer lie-ins, who seem to carry an increased risk of death from modern ailments, we might all wish we had inherited a little bit more from our early rising cousins.[343]

A nocturnal experience to savour

Beneath a moonless spring sky in a Sussex woodland, a 30-strong crowd slowly unknotted itself, revealing the string of nighthawks hidden within. Led by our experienced guide, we embarked cautiously yet purposefully into the midnight darkness. In accustoming our retinae to the dimness, we were noticeably reliant on our peripheral photoreceptors (the monotone rods), which are usually overlooked during daylight hours in favour of our central photoreceptors (the colourful cones). The skyward vista was starry, and Venus was setting directly ahead of us.

Every crevice, curve and contour of the woodland floor demanded a superlative effort from nervous systems usually so reliant on, but now largely bereft of, visual inputs. If a break emerged in the line, a gentle whistle of *woo-woo-woo-woo* from the group adrift would signal to our guide to pause. An echo of *woo-wee-woo* from the back to the front would indicate our reconnection, and our slow, silent march in the dark would recommence.

Meanwhile, the trees fizzed the most delicate pitter-patter. Perceived more audibly owing to the lack of illumination, the trees were being eaten alive. Insects gnawed, chiselled and gouged the bark and leaves, producing a continuous sprinkle of organic debris, which powdered the earth beneath. This so-called frass was what we could hear in the stillness of the night air, playing an important ecological role in cycling carbon and nutrients through the woodland habitat. Nature's gold dust was busy breaking her own golden silence.

But our mission that night was to experience a very different yet equally magical sound of Nature. It is one that has enchanted European cultures for millennia, but sadly one that's now confined in the UK to only fragmented pockets of activity. The performer of this sound is the national bird of Ukraine, where it's considered a symbol of love and fidelity. This vocalist sings 'of summer in full-throated ease' among 'shadows numberless', eulogised the English poet John Keats. And according to Greek mythology, the mutilated, tongueless Philomela only regained her song through her metamorphosis into this evocative avian singer.

Nevertheless, a successful mission was not guaranteed, for we were seeking Nature at its most vulnerable. According to the latest records from the British Trust for Ornithology, this bird species' current breeding population in England has declined to just 10% of its former 1965 levels. Alongside notable contractions in distribution across the island, the currently estimated 5,550 breeding pairs have been red-listed as a UK Bird of Conservation Concern since 2015.[344]

The object of our attention that evening was the Common Nightingale *Luscinia megarhynchos*. And sadly, it is a species in ecological trouble.

As a migratory bird that spends the Northern Hemisphere winter in Sub-Saharan Africa, the Common Nightingale travels back to European breeding grounds in the spring. For a period of six to eight weeks in April–June, the males are tireless attractors of female mates between the hours of 11 p.m. and 5 a.m. And it was one of these per-formances that our group hoped to tune into during our midnight foray through the woods.

Our guide was Sam Lee, a folk singer, naturalist and nightingale authority. He led us to the site of the previous night's recital, hoping to encounter the same nightingale's melody. Once located, we each found a spot on the ground and laid out our blankets. Although it was paramount we didn't encroach too closely, for fear of disturbing the nightingale's mating ritual, the song was clear enough in the crisp rural air to be heard for up to a mile around. Before long, the rasping croaks of a nearby frog had even joined in, followed soon after by the dulcet notes of Sam Lee himself, as he masterly duetted with the singing bird. As I lay there looking up to the cloud-free darkness above, I could have easily convinced myself I was being serenaded by the stars themselves.

A key component of this magical experience was the heightened level of engagement it demanded from every one of us. None of Sam Lee's guests was accustomed to night-time jaunts through an unknown woodland with strict prohibitions on the use of modern electronic devices. Nonetheless, there was beguiling appeal in switching our sensory emphasis to a more auditory one in this way. For me, this experience was just as much about the sylvan silence as it was about the final concert itself.

So do we all need to go traipsing off into the woods at night to re-engage with Nature? No, of course not. In fact, that would be rather disruptive and frankly counter-productive. Most of the time, we can choose other ways to reconnect with Nature, always ensuring we maintain a respectful distance.

But what distance is respectful enough?

Why we sometimes need to stay away

In E.O. Wilson's final book, *Half-Earth*, he proposes a deceptively simple fix to the climate and biodiversity crises. All that is needed, he argues, is to split the planet's land and ocean territories into two: half for humans and half for the rest of Nature to flourish without interference. If achieved, biodiversity loss would plateau at a tolerable level, and at least 85% of species would survive. Nature's refuge would work for the benefit of all life in the only way it knows, filtering our skies, cleansing our soils, replenishing our oceans and restoring global ecological health.

Fortunately, this is similar in principle to the agreement set out by the 2022 Kunming-Montreal Global Biodiversity Framework, albeit targeting 30% for Nature instead of Wilson's suggestion of 50%. But implicit to such an approach is a rather confusing realisation. If the *Half-Earth* idea is ultimately about excluding ourselves from thriving natural hotspots, could it really be the case that humans need to *disconnect* from Nature rather than *reconnect*? At first glance, this seems to conflict with what I've been endorsing all along. As I attempt to unravel this conundrum, let me take you on a mindful detour.

To live vicariously is to live through the actions and feelings of others. If you have ever tried to do this explicitly, you may, like me, have found it a challenging thing to do. There is just something about someone else's tales of lavish holidays and flashy purchases that quickly converts the most vicarious of intentions into green-eyed envy. This is amplified by our daily interactions with social media, as it can often appear that many others are having a better time than we are. So why is the art of vicarious living so hard? Has it always been this way for our species, or is it a modern phenomenon?

Breaking it down, I figure there are two overlapping components to mastering the vicarious experience. These are empathy and imagination. To feel genuine empathy for someone else is challenging for the simple reason that it first forces you to block out everything your own brain is telling you about your current actions and feelings.

Only then do you acquire the brain capacity to even consider those of another individual. This draws parallels with what I said in Chapter 8 about our natural inability to evaluate the climate and biodiversity crises from afar. When all is said and done, to put ourselves in the shoes of others requires an immense cognitive effort.

But that's not to say it can't be done. Like most things in life, it requires practice inspired by a willingness to achieve the desired outcome. Some people naturally find this task easier than others, based on their genetic heritage and cultural upbringing, but I would argue that everyone is capable of some improvement.

In my job as a doctor, I witness variable levels of suffering on an almost daily basis. While I don't experience that suffering myself, I try to empathise with my patients as fully as my brain will permit. Clearly, though, there is a caveat. How can I fully empathise with someone who is suffering in ways I have never personally experienced? I don't know what it feels like to have motor neuron disease, multiple sclerosis or Parkinson's disease, simply because I have never been diagnosed with these conditions. All I can do is imagine.

Even if you have never considered yourself a very imaginative person, do not discount its power. When it comes to imagination, humans are contenders for being Nature's most adept.

Our movement pathways imagine their intended task and adjust themselves before sending the final message out to muscles. Our dreams are an imagined, distorted portrayal of recent events, as our brains upgrade their memory stores. Hallucinations are the result of an overzealous imagination in visual brain areas. And when I froze in front of a larger-than-usual audience while competitively summarising my PhD, my brain's fear response was triggered by an imagined threat.

But going beyond these looser interpretations of what imagination can be, we spend most, if not all, of our waking lives imagining. In every conversation we have, in every news article we read, in every film we watch, in every work report we write, in every odour we smell, we rely on our imaginative powers. When we hear the word 'green' or read the word 'forest', it conjures up specific meanings and corresponding images in our minds. When we see a film hero or heroine admirably feigning their distress or delight on screen, we are even capable of re-imagining the already imagined. Just try to think about nothing, and something invariably pops up in one's mind.

The imagined world that exists in every human consciousness is no accident. As I discussed in Chapter 5, this is the tool that permits a workable understanding of the world around us. We embellish and fabricate the stories we tell each other, because to do otherwise would be like a tree trying to photosynthesise without its leaves.

The conclusion I draw from this is that any difficulty we have in living vicariously is not because of deficient imaginative powers, but because of inadequate empathic capacity. Furthermore, my proposition is that our difficulty to liberate that capacity is a direct consequence of our modern detachment from Nature. When you regularly surround yourself with the restorative, calming effects of the natural world, you feed the ancient biophilic circuits in your brain. In each birdsong you smile at, in each butterfly you feel the flutter of, in each flower petal you revel in, you find something that opens up the empathic channels in your mind, boosting your capacity for vicarious living.

And by revitalising this natural connection on a regular basis, it reinforces the inseparable bond our brains have forged with every other living being. When this innate bond is felt consciously, and not simply cast aside into our subconscious as modern society so often forces us to do, you have all the empathic capacity and imaginative powers you need for any vicarious experience you choose to master.

So the answer to the *Half-Earth* conundrum is simple. The physical disconnection imposed by this approach does not conflict with our reconnection with the natural world at all. In fact, it is a beautiful endorsement of it. By first physically reconnecting with the natural world on our doorsteps, we foster our psychological union with the wonder and intrigue it contains. It follows that the most natural response is to enhance the lives of as many trees, as many moths, as many birds and as many humans as you can, wherever they may be on Earth. If that means, as Wilson proposed, restricting our species to half the planet's territories, then so be it.

Translating our imaginations into practice

As I approach the end of this book, I want to include some practical guidance. What steps can we feasibly take to improve the unseen knock-on effects of our actions and promote the widespread conservation of health? This isn't about placing undue blame or expectation on individual choices but realising that small resets to our lifestyles can cumulatively turn the tide.

There are many excellent books on this topic, which offer a comprehensive, pragmatic overview of climate change and our influential role as denizens of this planet. Believe it or not, I never set out for this to be a book about climate change per se, but it turns out it is impossible to disentangle the Nature–health connection from the most important existential threat currently facing life on this planet. Therefore, let me first highlight the most valuable lessons I have learned from others, before adding several points of my own.

It seems appropriate to return to our nightingale navigator, Sam Lee, who interspersed his starstruck celebration of the Common Nightingale with some key messages. In the final chapter of his accompanying book, *The Nightingale*, he encourages everyone to eat less meat, be more charitable, buy fewer new clothes and switch to a green-minded bank.[345]

I must admit that this final one intrigued me. And it took many months before I eventually switched my 20-year-old HSBC account to a Nationwide one. The proverbial nail in the coffin of my old account arrived in the form of an excellent resource called Switch It Green. What I learnt was that even an account with nothing in it is used as part of a bank's customer base in order to secure its loans and assets. Arguably, if your bank is one that knowingly invests billions of dollars into destabilising our planet's climate and ecosystems, then the single most beneficial thing you can do is take your money out and close your account. If you do decide to do this, be sure to tell your old bank why.[346]

In her book *Saving Us*, Katharine Hayhoe emphasises the importance of talking. To extend the Carrollian metaphor from Chapter 8, perhaps this approach could be summarised as *solvitur loquendo* (it is solved by talking). We need to be talking about climate change, the biodiversity crisis and our uncertain future as often as we can. But this should not be a didactic exercise, since few people, especially those closest to us, respond well to being lectured.[347]

Instead, Hayhoe recommends approaching this seemingly impenetrable topic from an area of shared understanding. This involves identifying a common interest or mutual belief, however tangential it might seem, and somehow linking it to the critical climatic changes that our planet is experiencing. This is unlikely to change viewpoints overnight, but it has the best chance of sowing the seeds of reputable knowledge that will grow and flourish over time from within.

If you have an uncle passionate about golf, you could talk about how a land area four times bigger than all the golf courses in the world is being deforested across the globe every year. While the majority of this deforested land is said to be offset by reforestation, the new arboreal monocultures are no rivals for the richly biodiverse carbon sinks they often replace. As your uncle mulls it over, he might even think about the wider ecological impact of all those pristine golf courses.

Or, if you have a friend running their own social media channel, you could talk about how the centralised storage of over 100 zettabytes of cloud data (1 zettabyte = 1,000,000,000,000 gigabytes) across a rapidly increasing number of power-hungry, water-thirsty data hubs worldwide contributes around half the carbon footprint that the entire global healthcare system does. As we all consider this, we might even modify how we compile, edit and store our own digital content.

Next up is Bill McGuire's *Hothouse Earth*. But there is so much crammed into this concise book that it is challenging to select the most important advice. He mentions some things we can all do, such as choosing to walk or cycle over shorter distances, using trains instead of planes over longer distances and voting for the politicians with the greenest credentials. But he also clamours for change that seemingly goes beyond what an individual can realistically influence within the prevailing economic system. Like many others, McGuire demands an improved metric of worldly progress.[348]

The problem is that capitalism's insistence on limitless growth in a resource-limited natural world can only end one way: capitulation. The very existence of Earth Overshoot Day is an annual reminder of our civilisation's unsustainability. Any 'success' based on current capitalist measures, such as gross domestic product, is an illusion, since it pays insufficient heed to the destruction of the natural wealth that created it. This is a natural wealth with biodiversity and biophilia as the basic forms of currency. Until such measures are routinely brought into fiscal calculations, then we exhibit no greater maturity than kids stealing from a sweet shop.[349]

Indeed, the unsustainability of the capitalist model is the core message of a 2021 government report led by Partha Dasgupta, entitled *The Economics of Biodiversity*. Depending on how eager you are, you can peruse the 610-page full report, the 103-page abridged version or the much more digestible 10-page headline messages. Helpfully, all versions are free on the UK government website.[350]

I credit this report (the abridged version at least) for initiating the cognitive cogs that stirred this book into being. When I read that our economic models are purposefully designed to not only ignore the loss and contamination of the natural world, but also to treat human activity as external to Nature, I started to see the world through a different lens. The fact that the social and economic productivity of Nature is often 'silent, invisible and mobile' makes it so easy to snub. But despite the initial clarity of this new lens, the deeper I dived, the murkier it all became.

It just wasn't clear to me how I could translate the rational eloquence of this report into my daily decision-making. Given we are all forced to operate within this system, is it even possible to make any meaningful change? There are just so many variations of greenwashing to trip us up, such as greencrowding, greenrinsing, greenlighting, greenhushing, greenshifting and greenlabelling. These are deliberate strategies employed by corporations profiting from the capitalist agenda to dawdle, vacillate, deceive, cower, scapegoat and distort with respect to their ecological responsibilities. Thankfully, there are those fighting back.

Set up in 2006 on the premise that there is no planet B for us to move to, B Lab Global drives an international network of organisations to shift us towards 'an inclusive, equitable, and regenerative economy'. Businesses can assess their societal and ecological impact and adjust their behaviour in order to adhere to the rigorous standards set. And with overlapping visions of the future, the Rainforest Alliance's own certification programme similarly recognises those companies engaging in 'social, economic, and environmental sustainability'. Collectively, as consumers, we can send a powerful message by looking out for the B Corp and Rainforest Alliance logos on the items we purchase.[351]

A remedy that resets the baseline

Having picked out some valuable advice from other sources, I will now share three practical recommendations based on the contents of this book.

The first is *to renaturalise your senses on a daily basis*. This might be in a garden, a town park or an Area of Outstanding Natural Beauty. It might be for five minutes or five hours at a time. When we do not make a conscious effort to do this, it is so easy, even *expected* of us, to spend most of our lives indoors. Even when we do go outside, we are so often immersed in the least natural of surroundings, particularly for the growing majority of urban dwellers worldwide, that our senses must experience what a claustrophobic fears in enclosed spaces. Renaturalising your senses is about giving your visual, auditory, olfactory, gustatory and tactile pathways the regular space they need to engage your biophilic networks. If you don't feel better for it, then please write to me and let me know.

My second piece of practical advice is *to resist the division that engulfs diversity*. This means striving for the balance, compromise and respect required for measured debate. Whenever you detect the rocky edges of dualism in a conversation, retreat to more even ground. Whenever you hear the loudest proclaim with little regard for others, withdraw to quieter corners. Even when you read something you agree with, assess the language, tone and style used and ask yourself whether it has been written from a place of balance, compromise and respect. If not, it may just end up making matters worse. The reason this is so important is because without diversity Nature simply breaks down. All the adaptations and variations we observe across the biosphere, including within human society, are not only the *result* of a thriving, evolving process that has grown since life's conception, but also the *prerequisite* for its healthy continuation. In my view, it must be conserved at all costs.

My third takeaway is *to reimagine what makes you happy*. But before you do this, I urge you to strip the fleeting influence of modernity away. Since both your imagination and happiness are fundamental outputs of your brain's enduring biochemistry, not only are they influenced by the same natural inputs that have sustained those biochemical circuits since life began, but they are also equally contingent on one another. Let your inner creativity re-wild itself, and your inner sense of satisfaction will blossom. It matters little whether your chosen tool is a mic or a microscope, a pen or a pennywhistle, as these items are just the contrivances of modern expression. What really matters is what's been going on inside the biochemical hominin brain since who knows when.

In wrapping this section up, I want to return to the importance of small resets in the context of any practical steps we take in life. You may have already come across the concept of the shifting baseline syndrome, which describes the rolling renormalisation and deepening amnesia of any deteriorating natural condition. To illustrate this, there is an excellent cartoon depiction of a subsurface seascape. While the panel from the year 1800 shows a healthy variety of marine plants and animals, the panel from 1950 displays far fewer. But it's the 2019 panel, with its exchange of ocean life for plastic bags, that demonstrates how far the baseline has shifted across two centuries. From one generation to the next, the meaning of 'normal' continually morphs.[352]

As depressing as this sounds, it does give cause for hope. If the baseline can shift one way, why can't it be shifted back? If so, then it will more than likely require lots of small resets along the way. Like the seasonally shifting Eurasian sunrises and sunsets, which continue to provide the same circadian resets today as they did in the past, small resets to our lifestyles have a much greater chance of making a sustainable impact. If you can eat 20% less beef, buy 20% fewer clothes and cut out one return flight each year, then this is much better than trying to eliminate these activities completely and finding it too tricky to sustain. Of course, once you've got used to the new and improved 'normal', you can make another favourable reset.

I call this the resetting baseline remedy.

Living up to our name

I now want to share something so elegant, something so admired across the centuries, that it will be difficult to do it justice. It can be entirely expressed by three letters, two numbers and two mathematical symbols. It has been described as 'the most profound mathematical statement ever written'. In *Mathematics and the Imagination* by Edward Kasner and James R. Newman, it is said to 'appeal equally to the mystic, the scientist, the philosopher and the mathematician'. As such, it is widely considered to be the most beautiful equation in the world. This equation is known as Euler's Identity.[353]

Leonhard Euler was an eighteenth-century Swiss mathematician, who on his death in 1783 left quite the legacy. Living in an Age of Enlightenment, Euler put his mind to a broad range of topics in physics, astronomy and engineering. He was particularly fascinated by prime numbers and the three-body problem, which concerns how the Sun, Earth and Moon move in relation to one another. Testament to his insatiable work ethic, he considered blindness at the age of 59 as simply one less source of distraction.[354]

So what is Euler's Identity and what makes it so special? First, here is the famous equation:

$$e^{i\pi} + 1 = 0$$

To break it down, let's explore its constituents in turn. I will begin with a number that should be familiar from maths lessons growing up. The Greek letter that represents this number is π (*pi*; pronounced the same as the English word 'pie'). Formally, it is the ratio of a circle's circumference to its diameter. It approximates to 3.14159, but to write down its exact value would require an infinite quantity of paper. This is because the digits to the right of the decimal place go on for ever in a non-repetitive fashion. This property makes π an irrational number.

The next letter I will explain denotes another irrational number. It is written as **e** and it approximates to 2.71828. Its uniqueness stems from its role in mathematics as the natural base. This can be appreciated visually in many natural contexts, including a mollusc's swirling shell, the formation of a cyclone, the seed arrangement on a sunflower head and the flight paths of hunting falcons. All these natural entities produce spirals that are invariant to scale, reminiscent of the fractal patterns I discussed in Chapter 7. And the self-similarity of these natural spirals only exists because of the magical, mathematical properties of **e**.[355]

The final letter to introduce is *i*. This represents not an irrational number like π or **e**, but an imaginary one instead; *i* is the number you get when you take the square root of –1. As it has no physical representation, it can only be imagined.

The two remaining numbers in Euler's Identity are **0** and **1**, the binary workhorses of both digital and natural systems, as we saw in Chapter 5. While these two numbers may feel more intuitive than the irrational or imaginary entities of π, **e** and *i*, the concepts of zeroness and oneness are not so clear cut.

To think of one apple or one car seems simple enough, but what about one day? As one day is equal to 24 hours, 1,440 minutes or 86,400 seconds, shouldn't this make us question what oneness really means? Its inherent divisibility seems to pose a problem. Perhaps, then, we should think of oneness as the absence of zeroness. But if we do that, it brings us onto what zeroness means.

It is tempting to equate zeroness to nothingness or emptiness. But do these concepts really exist? While it is true that an empty fruit bowl has no apples in it and an empty drive has no cars on it, there are still, at the very least, microbes on their surfaces and gas molecules in the air above them. So maybe a vacuum in space represents true nothingness or emptiness. But even space has measurable dimensions and allows light to pass through it.

Maybe, then, we need to think about zeroness differently. Let's consider the empty fruit bowl again and compare it with a second fruit bowl containing one apple. What happens when we put the contents of the empty fruit bowl into the one with a single apple? Of course, we still have one apple. In other words, the addition of zero apples makes no change to the original amount. Therefore, isn't zeroness a better representation of stability and consistency rather than nothingness and emptiness?

The reason I am labouring these interpretations is because it influences how we might reflect on Euler's Identity more broadly. But given the frankly bizarre properties of these five numbers, it is staggering that Euler's Identity, in all its simplicity, is even true (see Figure 11 in the Appendix for a brief explanation). There is no doubt that its derivation in the eighteenth century triggered an ardent search for some deeper, almost spiritual significance. Why do five of the most fundamental numbers in mathematics link together so neatly?

Almost three centuries on, permit me to put my own spin on it.

e is the Veery singing its duet in the Canadian afternoon as well as the Common Nightingale riffing in the British midnight. e is the Parent Bug protecting its nymphs up close as well as the Dark-edged Bee-fly ejecting its eggs from afar. e is the fast-moving Green Anole lizard adapting to a freak Texan freeze as well as the fast-shifting crystalline complexion of the Common Chameleon. e is the mutable home of a gut microbiome as well as the mutating tome of a human genome. e is ring speciation, brood parasitism and biophilia. e is all the strands of Nature in one.

But as Euler's Identity shows, e does not operate alone. It relies on an inseparable union between i and π, symbolising the imagined and sensed aspects of our natural world.

i is in every mind that pictures a bird from its song alone. *i* is in every parental deliberation that raises the next generation. *i* is in every story of humanity's mortal lot. *i* is in every dream, every category and every treatment that Nature inspires. Meanwhile, π is what excites auditory pathways in the springtime woodland. π is the nymphs' protective shield and the parasitised bee's hole in the ground. π is the glimmering sea of guanine nanocrystals in a chameleon's skin. And it is only when e, *i* and π combine that the full natural experience is complete.

Every one of these experiences then triggers a balanced reaction. This could come in the form of a biochemical adaptation that leads to a tweak in behaviour, or it could just strengthen a connection that already existed. Whatever the 1 in Euler's Identity represents in practice, its sole purpose is to provide a route back to the stability of 0. On one level, nothing changes. And yet, deeper down, everything does. In that way, Euler's Identity is just one more of Nature's infinite spirals of sustainability.

It seems apt that only a decade separates Euler's unveiling of his 'identity' in 1748 and Carl Linnaeus' labelling of our species as the 'wise ape' in 1758. But as we reflect on our influence on the natural world to date, perhaps the wisest thing we can do is to ditch such a lofty self-assessment. Maybe a revised title, *Homo insapiens*, meaning the 'unwise ape', would cast a more grounding influence on us all. Or maybe *Homo insapiens* will lie in wait until Earth's next wise species renames us in our wake.

A final thought from me

In concluding, I want to return to the human disease that has formed the focus of my medical research. Motor neuron disease is terrifying. It silences you, paralyses you and chokes you. It does all of this from within, unseen and unheard. However, it also tends to spare the machinery that keeps you thinking and strategising and resisting. The result is a desperate form of entrapment that is continuously striving to break free.

Mercifully, what we're witnessing now in the battle against motor neuron disease is a tangible chink in its armour. For a small group of patients with specific DNA mutations, a new therapy is conserving the health of motor neurons otherwise destined for demise. All that silencing, paralysing and choking is being held back. Akin to the therapeutic breakthrough for children with spinal muscular atrophy, the dream of a cure is tentatively stepping out of the shadows.

But complex problems take time to solve. Arguably, there is no problem more complex than our species' deteriorating relationship with Nature. Our task is to reawaken that inner connection, as if it's just been dozing a while. We must use all the thinking and strategising and resisting we can muster. That way, we will conserve whatever health we can, and all of Nature may flourish once again.

Epilogue

About a month after I had finished writing the last chapter, I met a new species for the first time. But given the countless species across life's kingdoms that operate beyond our usual range of familiarity, this is hardly newsworthy; and almost certainly not enough to base an epilogue on.

However, there *was* something extraordinary about this meeting. And the more I learnt about it, the more it seemed to unify the main themes from this book. Eight months on, I can no longer resist sharing it.

As is so often the case when engaging with the natural world, I didn't know the name of this new species at the time. You'll remember that the same thing happened with the Veery, whose anonymous song captivated me in a Canadian woodland. And it happened this time too, albeit much closer to home.

It was 9 May 2024. The sun was out, and temperatures were on the rise. As I walked my dog Milo in the local woods, I remained as vigilant as ever, seeking any springtime birds, flowers or invertebrates that might catch my eye. This is a patient pastime, something that Milo regularly fails to grasp, as he waits and whimpers up ahead. But ready with camera in hand, I was glad to find several woodland residents to photograph that day. So, later that evening, I did as I'm accustomed to do. I downloaded the pictures onto my phone and sent the ones of interest to my friend, David Cousins.

David is a lifelong birder and an avid moth-er, most recently becoming a full-blown pan-species lister. Whether it concerns birds, invertebrates, plants or any of Nature's wares, he knows far more about species identification than I ever will. And I frequently lean on his expertise.

'Looks odd,' replied David, after I sent him a picture of what looked to me like a fairly regular shield bug. It was dark brown with orange and black antennae, and it had stripy rims on each side of its abdomen. The picture was sadly not as crisp as I would have liked. And given the dense canopy overhead, I had needed to brighten the photo quite a bit in the edit. In the grand scheme of things, I hadn't been overly excited by it.

'I'm having it checked with a mate,' David added.

David doesn't usually need to check with a mate. I was intrigued. Had I captured some rare mutant? Or was it a vagrant that had flown in from a far-flung shore?

David asked if I had taken the picture in Essex.

'Yes,' I answered.

And I keenly waited for his reply.

To understand why David was less familiar with this shield bug than all the other insects I have ever sent him, I must explain something known as an indicator species.

This is a species that tells scientists and conservationists something critical about the environment in which it lives. It can include the Red-legged Honeycreeper *Cyanerpes cyaneus* of Belize, whose preference for the forest edge makes it an indicator of human disturbance. It can include a particular species of grass skipper butterfly, known as the Sachem *Atalopedes campestris*, whose recent advance up the West Coast of the USA has been facilitated by warming winters. Or it can include another species of butterfly, the De Prunner's Ringlet *Erebia triaria*, which has been pushed to higher altitudes in Spain's Sierra de Guadarrama mountains in order to escape the creeping heat from below.[356]

Whether it's the emergence of a new species in one area or the loss of an established species elsewhere, indicator species spell change. The underlying drivers might include pollution, food shortages and changing climates. But as habitats change, the fussiest species move on to maintain their ecological niche. And this gives scientists a shortcut for understanding the stability and, by extension, the health of a natural area.

So why is this relevant to the shield bug I met on a pleasant spring day in Essex, UK? And does this explain why David had to seek a second opinion?

Thankfully, David didn't keep me waiting long. Just 17 minutes after sending him my photo, I received confirmation that the species I had photographed was an insect known as the Vernal Shieldbug *Peribalus strictus*. And it had created quite the stir among a local network of enthusiasts. But why had this observation generated such interest in such a short space of time?

The answer was simple. This was the first time this species had *ever* been recorded in North Essex.

To find out more, I reached out to Tristan Bantock, Consultant Entomologist and co-founder of the British Bugs website. 'It is not a species I have seen in Britain,' confirmed Tristan, after sending me a map of UK sightings. With only a sprinkling of records pre-2000 and nearly all subsequent appearances occurring along the south coast of England, this is a species that has been steadily moving northwards as mainland Europe warms. The Vernal Shieldbug, therefore, is an indicator species of climate change.

As surprising as this was, I found myself in two minds. Should I be celebrating this encounter? Or should I be lamenting it?

After all, observations of this kind present a dilemma to naturalists, even amateur ones such as myself. On the one hand, it is exciting to see new species populate your local area. But on the other, one cannot ignore the drivers of these worrying population shifts.

I imagined what would happen if the Vernal Shieldbug were forced to continue its northward advance over the following decades. One day, will it even find Scottish temperatures unbearable? If so, where will it go then? Equally, once the De Prunner's Ringlet from the Sierra de Guadarrama runs out of mountaintop, where will it go next?

These species are currently able to stay one step ahead of a changing climate, because there is suitable neighbouring habitat in reserve for them. But if they continue to get pushed to the extremes, one day they will simply have nowhere to go.

For this reason, it is imperative that we do not bury our heads in the sand. We need many more observations like the one I shared with David and Tristan. This will then allow the experts to glean what they can from a whole range of indicator species, including the Vernal Shieldbug. Thankfully, there is an excellent resource to help. It is called iRecord, and it is where I have logged my sighting of the Vernal Shieldbug. At the time of writing, over 21.5 million records of around 24,000 species have been submitted by 220,000 people. This is a phenomenally large databank that we can all contribute to.[357]

And as it turns out, it's the smaller species, such as the Vernal Shieldbug, that are likely to be the most informative. This is because insects make excellent indicators of climate change. They not only shift their habitat ranges easily but can also readily adapt to environmental pressures over short time frames. They can switch on heat shock proteins, adapt the size and colouration of their body parts, shorten the interval between reproductive cycles and alter their daily patterns of behaviour. And all these adaptations, plus many more, work together so that certain species are able to survive the challenges posed by a rapidly changing planet Earth.[358]

But as important as these observations are, I have to be honest with you. I don't walk in Nature to find indicator species. I walk in Nature because it makes me feel better. It is my outlet from a full-on work schedule and a busy home life. It relieves stress and restores focus. And if I happen to meet an indicator species or two along the way, then that is a valued bonus.

As it is for so many, walking in Nature is my flexible, cheap and fulfilling health regime. After all, the natural world is a healthy place to be. A deep breath of fresh woodland air just *smells* healthy. A jaunty song from the treetops pulsating my eardrums just *sounds* healthy. A flap of marbled monotones settling on a radiant dandelion flower just *looks* healthy. And while my camera enriches my engagement with Nature for the purposes of sharing and recollecting, it is a technological artifice that must remain sympathetic to both the natural environment and my own brain's need for direct interaction.

But the regular engagement that any of us have with the natural world thankfully goes way beyond what we might feel as individuals, even if that's almost always where it starts. It is leading to improved knowledge all the time, about the mechanisms of Nature connectedness, about the importance of indicator species, about the benefits of biodiversity, about the perils of a rapidly changing climate and about how we can best conserve the health of our natural world.

For Nature is on our doorsteps. It is there because we are part of it too. Nature doesn't leave us alone for the same reasons that we cannot abandon it either. Nature runs through our genes, our brains and our blood. It continues to shape our minds, bodies and health as it always has done. Nature is each factory of energy within our cells that allows us to think, feel, act, imagine and do. But for too long, modern *Homo sapiens* has been slamming the door in Nature's face. Collectively, though, this *can* change.

So go forth. Renaturalise your senses, and feel happier and healthier as a result. Then, go and naturalise them some more. In this complex, somewhat confusing world, it may be about the only thing that still makes perfect sense.

Appendix

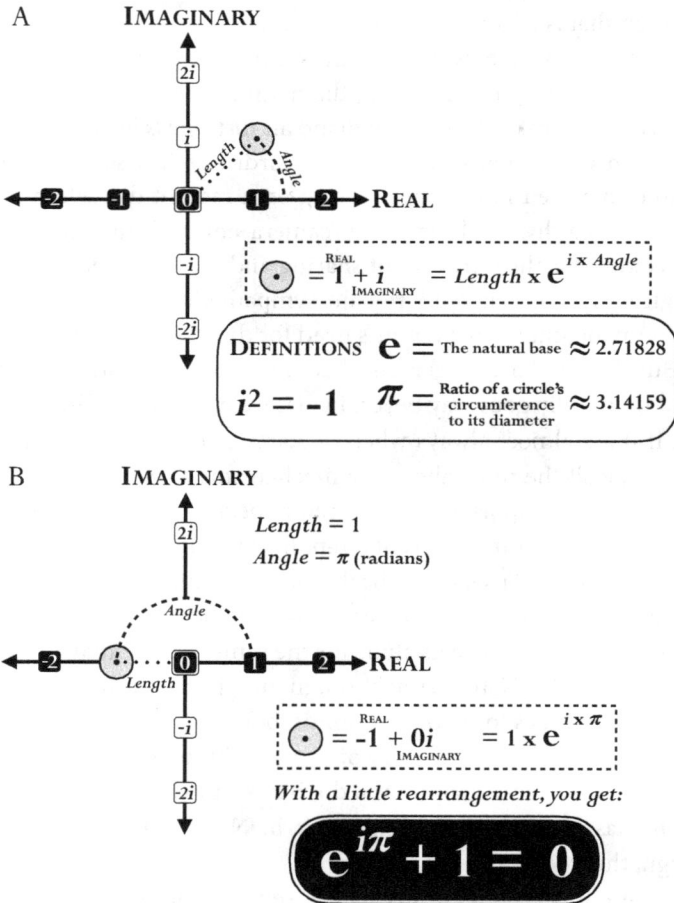

Figure 11. Euler's Identity: A geometrical explanation of the formula that links five of the most fundamental numbers in mathematics. A. General case using 1 + i as an example; B. Specific case that leads to Euler's Identity.

References

Introduction

1. Claessens, H., Oosterbaan, A., Savill, P. and Rondeux, J., 2010. A review of the characteristics of black alder (*Alnus glutinosa* (L.) Gaertn.) and their implications for silvicultural practices. *Forestry: An International Journal of Forest Research*, 83(2), pp.163–175. https://doi.org/10.1093/forestry/cpp038

2. Alloisio, N., Queiroux, C., Fournier, P., Pujic, P., Normand, P., Vallenet, D., Médigue, C., Yamaura, M., Kakoi, K. and Kucho, K., 2010. The *Frankia alni* symbiotic transcriptome. *Molecular Plant-Microbe Interactions*, 23(5), pp.593–607. https://doi.org/10.1094/MPMI-23-5-0593

3. Brusatte, S.L., O'Connor, J.K. and Jarvis, E.D., 2015. The origin and diversification of birds. *Current Biology*, 25(11), pp.R888–R898. https://doi.org/10.1016/j.cub.2015.08.003

4. Ernster, L. and Schatz, G., 1981. Mitochondria: a historical review. *Journal of Cell Biology*, 91(3), pp.227s-255s. https://doi.org/10.1083/jcb.91.3.227s

5. Dunn, J. and Grider, M.H., 2023. Physiology, Adenosine Triphosphate. *StatPearls*. Treasure Island, FL: StatPearls Publishing LLC.

6. Martin, W.F., Garg, S. and Zimorski, V., 2015. Endosymbiotic theories for eukaryote origin. *Philosophical Transactions of the Royal Society B: Biological Sciences*, 370(1678), p.20140330. https://doi.org/10.1098/rstb.2014.0330

7. Sender, R., Fuchs, S. and Milo, R., 2016. Revised estimates for the number of human and bacterial cells in the body. *PLoS Biology*, 14(8), p.e1002533. https://doi.org/10.1371/journal.pbio.1002533; Bannert, N. and Kurth, R., 2006. The evolutionary dynamics of human endogenous retroviral families. *Annual Review of Genomics and Human Genetics*, 7, pp.149–173. https://doi.org/10.1146/annurev.genom.7.080505.115700

8. Salmon, J., 1951. Vegetable caterpillars. *Tuatara: Journal of the Biological Society*, New Zealand Electronic Text Collection, pp.1–3.

9. Heine, D., Holmes, N.A., Worsley, S.F., et al., 2018. Chemical warfare between leafcutter ant symbionts and a co-evolved pathogen. *Nature Communications*, 9(1), p.2208. https://doi.org/10.1038/s41467-018-04520-1

10. Ibid.

11. Earth Overshoot Day, 2025. Available at: https://www.overshootday.org

12. McInerney, F.A. and Wing, S.L., 2011. The Paleocene-Eocene Thermal Maximum: a perturbation of carbon cycle, climate, and biosphere with implications for the future. *Annual Review of Earth and Planetary Sciences*, 39, pp.489–516. https://doi.org/10.1146/annurev-earth-040610-133431.; Voosen, P., 2019. Project traces 500 million years of roller-coaster climate. *Science*, 364(6442), pp.716–717. https://doi.org/10.1126/science.364.6442.716

13. Bryson, B., 2004. *A Short History of Nearly Everything*. London: Crown.

Chapter 1

14. Kaplan, R. and Kaplan, S., 1989. *The Experience of Nature: A Psychological Perspective*. Cambridge: Cambridge University Press.

15. Ulrich, R.S., Simons, R.F., Losito, B.D., et al., 1991. Stress recovery during exposure to natural and urban environments. *Journal of Environmental Psychology*, 11(3), pp.201–230. https://doi.org/10.1016/S0272-4944(05)80184-7

16. Dosen, A.S. and Ostwald, M.J., 2016. Evidence for prospect-refuge theory: a meta-analysis of the findings of environmental preference research. *City, Territory and Architecture*, 3(1), p.4. https://doi.org/10.1186/s40410-016-0033-1

17. Gupta, M., Ireland, A.C. and Bordoni, B., 2023. Neuroanatomy, Visual Pathway. *StatPearls*. Treasure Island, FL: StatPearls Publishing LLC.

18. Schadlu, A.P., Schadlu, R. and Shepherd, J.B., 2009. Charles Bonnet syndrome: a review. *Current Opinion in Ophthalmology*, 20(3), pp.219–222. https://doi.org/10.1097/ICU.0b013e328329b643

19. Saionz, E.L., Busza, A. and Huxlin, K.R., 2022. Rehabilitation of visual perception in cortical blindness. *Handbook of Clinical Neurology*, 184, pp.357–373. https://doi.org/10.1016/B978-0-12-819410-2.00030-8

20. Bruss, D.M. and Shohet, J.A., 2023. Neuroanatomy, Ear. *StatPearls*. Treasure Island, FL: StatPearls Publishing LLC.

21. Maguire, M.J., 2012. Music and epilepsy: a critical review. *Epilepsia*, 53(6), pp.947–961. https://doi.org/10.1111/j.1528-1167.2012.03523.x; Basner, M., Babisch, W., Davis, A., et al., 2014. Auditory and non-auditory effects of noise on health. *The Lancet*, 383(9925), pp.1325–1332. https://doi.org/10.1016/S0140-6736(13)61613-X

22. Kirste, I., Nicola, Z., Kronenberg, G., et al., 2015. Is silence golden? Effects of auditory stimuli and their absence on adult hippocampal neurogenesis. *Brain Structure and Function*, 220(2), pp.1221–1228. https://doi.org/10.1007/s00429-013-0679-3; Voisin, J., Bidet-Caulet, A., Bertrand, O., et al., 2006. Listening in silence activates auditory areas: a functional magnetic resonance imaging study. *Journal of Neuroscience*, 26(1), pp.273–278. https://doi.org/10.1523/JNEUROSCI.2967-05.2006

23. Sharma, A., Kumar, R., Aier, I., et al., 2019. Sense of smell: structural, functional, mechanistic advancements and challenges in human olfactory research. *Current Neuropharmacology*, 17(9), pp.891–911. https://doi.org/10.2174/1570159X17666181206095626

24. Shaikh, F.H. and Soni, A., 2023. Physiology, Taste. *StatPearls*. Treasure Island, FL: StatPearls Publishing LLC.

25. Gadhvi, M., Moore, M.J. and Waseem, M., 2023. Physiology, Sensory System. *StatPearls*. Treasure Island, FL: StatPearls Publishing LLC.

26. Rolls, E.T., 2016. Reward systems in the brain and nutrition. *Annual Review of Nutrition*, 36, pp.435–470. https://doi.org/10.1146/annurev-nutr-071715-050725

27. Lord, C., Brugha, T.S., Charman, T., et al., 2020. Autism spectrum disorder. *Nature Reviews Disease Primers*, 6(1), p.5. https://doi.org/10.1038/s41572-019-0138-4

28. Schulz, S.E. and Stevenson, R.A., 2019. Sensory hypersensitivity predicts repetitive behaviours in autistic and typically-developing children. *Autism*, 23(4), pp.1028–1041. https://doi.org/10.1177/1362361318774559

29. Bliss, T.V. and Cooke, S.F., 2011. Long-term potentiation and long-term depression: a clinical perspective. *Clinics (São Paulo)*, 66 Suppl 1(Suppl 1), pp.3–17. https://doi.org/10.1590/s1807-59322011001300002

30. Daneman, R. and Prat, A., 2015. The blood-brain barrier. *Cold Spring Harbor Perspectives in Biology*, 7(1), p.a020412. https://doi.org/10.1101/cshperspect.a020412; Osadchiy, V., Martin, C.R. and Mayer, E.A., 2019. The gut-brain axis and the microbiome: mechanisms and clinical implications. *Clinical Gastroenterology and Hepatology*, 17(2), pp.322–332. https://doi.org/10.1016/j.cgh.2018.10.002; Tuthill, J.C. and Azim, E., 2018. Proprioception. *Current Biology*,

28(5),   pp.R194-R203.   https://doi.org/10.1016/j.cub.2018.01.064;   Casale,   J., Browne, T., Murray, I.V., et al., 2023. Physiology, Vestibular System. *StatPearls*. Treasure Island, FL: StatPearls Publishing LLC.; Jimsheleishvili, S. and Dididze, M., 2023. Neuroanatomy, Cerebellum. *StatPearls*. Treasure Island, FL: StatPearls Publishing LLC.

31.   Harvey, P.D., 2019. Domains of cognition and their assessment. *Dialogues in Clinical Neuroscience*,   21(3),   pp.227–237.   https://doi.org/10.31887/DCNS.2019.21.3/ pharvey

32.   LeBouef, T., Yaker, Z. and Whited, L., 2023. Physiology, Autonomic Nervous System. *StatPearls*. Treasure Island, FL: StatPearls Publishing LLC.

33.   Patton, A.P. and Hastings, M.H., 2018. The suprachiasmatic nucleus. *Current Biology*, 28(15), pp.R816-R822. https://doi.org/10.1016/j.cub.2018.06.052

34.   Groucutt, H.S., Petraglia, M.D., Bailey, G., et al., 2015. Rethinking the dispersal of *Homo sapiens* out of Africa. *Evolutionary Anthropology*, 24(4), pp.149–164. https://doi.org/10.1002/evan.21455

35.   Melrose, S., 2015. Seasonal affective disorder: an overview of assessment and treatment approaches. *Depression Research and Treatment*, 2015, p.178564. https://doi.org/10.1155/2015/178564

36.   Sintzel, M.B., Rametta, M. and Reder, A.T., 2018. Vitamin D and multiple sclerosis: a comprehensive review. *Neurological Therapy*, 7(1), pp.59–85. https:// doi.org/10.1007/s40120-017-0086-4

37.   Simpson, S. Jr., Blizzard, L., Otahal, P., et al., 2011. Latitude is significantly associated with the prevalence of multiple sclerosis: a meta-analysis. *Journal of Neurology, Neurosurgery & Psychiatry*, 82(10), pp.1132–1141. https://doi. org/10.1136/jnnp.2011.240432

38.   Brugler, M.R., Aguado, M.T., Tessler, M., et al., 2018. The transcriptome of the Bermuda fireworm *Odontosyllis enopla* (Annelida: Syllidae): a unique luciferase gene family and putative epitoky-related genes. *PLoS One*, 13(8), p.e0200944. https://doi.org/10.1371/journal.pone.0200944; Fischer, A. and Fischer, U., 1995. On the life-style and life-cycle of the luminescent polychaete *Odontosyllis enopla* (Annelida: Polychaeta). *Invertebrate Biology*, 114(3), pp.236–247. https://doi. org/10.2307/3226878

39.   Poehn, B., Krishnan, S., Zurl, M., et al., 2022. A Cryptochrome adopts distinct moon- and sunlight states and functions as sun- versus moonlight interpreter in monthly oscillator entrainment. *Nature Communications*, 13(1), p.5220. https:// doi.org/10.1038/s41467-022-32562-z

40.   Wehr, T.A., Giesen, H.A., Moul, D.E., et al., 1995. Suppression of men's responses to seasonal changes in day length by modern artificial lighting. *American Journal of Physiology*, 269(1 Pt 2), pp.R173-R178. https://doi.org/10.1152/ ajpregu.1995.269.1.R173; Helfrich-Förster, C., Monecke, S., Spiousas, I., et al., 2021. Women temporarily synchronize their menstrual cycles with the luminance and gravimetric cycles of the Moon. *Science Advances*, 7(5), p.eabe2781. https:// doi.org/10.1126/sciadv.abe1358; Menaker, W. and Menaker, A., 1959. Lunar periodicity in human reproduction: a likely unit of biological time. *American Journal of Obstetrics and Gynecology*, 77(4), pp.905–914. https://doi.org/10.1016/ S0002-9378(16)36803-X; Arliss, J.M., Kaplan, E.N. and Galvin, S.L., 2005. The effect of the lunar cycle on frequency of births and birth complications. *American Journal of Obstetrics and Gynecology*, 192(5), pp.1462–1464. https:// doi.org/10.1016/j.ajog.2004.12.034

41.   Benedict, C., Franklin, K.A., Bukhari, S., et al., 2022. Sex-specific association of the lunar cycle with sleep. *Science of the Total Environment*, 804, p.150222.   https://doi.org/10.1016/j.scitotenv.2021.150222;   Cajochen,   C., Altanay-Ekici, S., Münch, M., et al., 2013. Evidence that the lunar cycle

influences human sleep. *Current Biology*, 23(15), pp.1485–1488. https://doi.org/10.1016/j.cub.2013.06.029; Casiraghi, L., Spiousas, I., Dunster, G.P., et al., 2021. Moonstruck sleep: synchronization of human sleep with the moon cycle under field conditions. *Science Advances*, 7(5), p.eabe0782. https://doi.org/10.1126/sciadv.abe0465

42. Hopkins, S., 1989. Experimental Musical Instruments, "Whirly Instruments", Book 5, Edition 3, Page 59. Available at: https://archive.org/details/emi_archive/EMI_5_3_October1989/page/10/mode/2up; Miller, M., 2018. Music for Whirly Tubes. Available at: https://www.youtube.com/watch?v=zjlWFE_9fVo

43. Riede, T., Thomson, S.L., Titze, I.R., et al., 2019. The evolution of the syrinx: an acoustic theory. *PLoS Biology*, 17(2), p.e2006507. https://doi.org/10.1371/journal.pbio.2006507

44. Brown, R.E., Brain, J.D. and Wang, N., 1997. The avian respiratory system: a unique model for studies of respiratory toxicosis and for monitoring air quality. *Environmental Health Perspectives*, 105(2), pp.188–200. https://doi.org/10.1289/ehp.97105188

45. Deng, T., Lu, X., Wang, S., et al., 2021. An Oligocene giant rhino provides insights into *Paraceratherium* evolution. *Communications Biology*, 4(1), p.639. https://doi.org/10.1038/s42003-021-02170-6

46. Hinds, D.S. and Calder, W.A., 1971. Tracheal dead space in the respiration of birds. *Evolution*, 25(2), pp.429–440. https://doi.org/10.2307/2406936

47. Zhang, Z., 2016. Mechanics of human voice production and control. *Journal of the Acoustical Society of America*, 140(4), pp.2614. https://doi.org/10.1121/1.4964509

48. Podos, J. and Cohn-Haft, M., 2019. Extremely loud mating songs at close range in white bellbirds. *Current Biology*, 29(20), pp.R1068-R1069. https://doi.org/10.1016/j.cub.2019.09.028

49. Suthers, R.A., 1990. Contributions to birdsong from the left and right sides of the intact syrinx. *Nature*, 347(6292), pp.473–477. https://doi.org/10.1038/347473a0

50. Langille, J.H., 1885. Our birds in their haunts: a popular treatise on the birds of eastern North America. *The Auk*, 2(1), pp.91–94.

51. Hobson, K.A. and Kardynal, K.J., 2015. Western Veeries use an eastern shortest-distance pathway: new insights to migration routes and phenology using light-level geolocators. *The Auk*, 132(3), pp.540–550. https://doi.org/10.1642/AUK-14-260.1

52. Heckscher, C.M., 2018. A Nearctic-Neotropical migratory songbird's nesting phenology and clutch size are predictors of accumulated cyclone energy. *Scientific Reports*, 8(1), p.9899. https://doi.org/10.1038/s41598-018-28302-3; Butler, R.W., 2000. Stormy seas for some North American songbirds: are declines related to severe storms during migration? *The Auk*, 117(2), pp.518–522. https://doi.org/10.1642/0004-8038(2000)117[0518:SSFSNA]2.0.CO;2

Chapter 2

53. Oliver, S. and Duncan, S., 2019. Editorial: Looking through the Johari window. *Research for All*, pp.1–6.

54. Sender, R., Fuchs, S. and Milo, R., 2016. Revised estimates for the number of human and bacterial cells in the body. *PLoS Biology*, 14(8), p.e1002533. https://doi.org/10.1371/journal.pbio.1002533

55. Watson, J.D. and Crick, F.H., 1953. Molecular structure of nucleic acids; a structure for deoxyribose nucleic acid. *Nature*, 171(4356), pp.737–738. https://doi.org/10.1038/171737a0

56. Dahm, R., 2005. Friedrich Miescher and the discovery of DNA. *Developmental Biology*, 278(2), pp.274–288. https://doi.org/10.1016/j.ydbio.2004.11.028;

Abbott, S. and Fairbanks, D.J., 2016. Experiments on plant hybrids by Gregor Mendel. *Genetics*, 204(2), pp.407–422. https://doi.org/10.1534/genetics.116.195198; Levit, G.S. and Hossfeld, U., 2019. Ernst Haeckel in the history of biology. *Current Biology*, R1276–R1284. https://doi.org/10.1016/j.cub.2019.10.064

57. Miescher, F., 1897. *Die Histochemischen und Physiologischen Arbeiten.* Leipzig: F.C.W. Vogel.

58. Hall, K. and Sankaran, N., 2021. DNA translated: Friedrich Miescher's discovery of nuclein in its original context. *British Journal for the History of Science*, 54(1), pp.99–107. https://doi.org/10.1017/S000708742000062X; Venter, J.C., Adams, M.D., Myers, E.W., et al., 2001. The sequence of the human genome. *Science*, 291(5507), pp.1304–1351. https://doi.org/10.1126/science.1058040

59. Al-Chalabi, A. and Hardiman, O., 2013. The epidemiology of ALS: a conspiracy of genes, environment and time. *Nature Reviews Neurology*, 9(11), pp.617–628. https://doi.org/10.1038/nrneurol.2013.203

60. Neumann, M., Sampathu, D.M., Kwong, L.K., et al., 2006. Ubiquitinated TDP-43 in frontotemporal lobar degeneration and amyotrophic lateral sclerosis. *Science*, 314(5796), pp.130–133. https://doi.org/10.1126/science.1134108

61. Volk, A.E., Weishaupt, J.H., Andersen, P.M., et al., 2018. Current knowledge and recent insights into the genetic basis of amyotrophic lateral sclerosis. *Medical Genetics*, 30(2), pp.252–258. https://doi.org/10.1007/s11825-018-0185-3

62. Bensimon, G., Lacomblez, L. and Meininger, V., 1994. A controlled trial of riluzole in amyotrophic lateral sclerosis. *The New England Journal of Medicine*, 330(9), pp.585–591. https://doi.org/10.1056/NEJM199403033300901; Blair, H.A., 2023. Tofersen: first approval. *Drugs*, 83(11), pp.1039–1043. https://doi.org/10.1007/s40265-023-01904-6

63. Hall, K. and Sankaran, N., 2021. DNA translated: Friedrich Miescher's discovery of nuclein in its original context. *British Journal for the History of Science*, 54(1), pp.99–107. https://doi.org/10.1017/S000708742000062X

64. Holesh, J.E., Aslam, S. and Martin, A., 2023. Physiology, Carbohydrates. *StatPearls.* Treasure Island, FL: StatPearls Publishing LLC.; Forest Research, 2023. Emissions. Available at: https://www.forestresearch.gov.uk/tools-and-resources/fthr/biomass-energy-resources/technical-and-regulatory/emissions/

65. Martel, V., Morin, O., Monckton, S.K., et al., 2022. Elm zigzag sawfly, *Aproceros leucopoda* (Hymenoptera: Argidae), recorded for the first time in North America through community science. *The Canadian Entomologist*, 154(1), e1. https://doi.org/10.4039/tce.2021.44

66. Martín, J.A., Sobrino-Plata, J., Rodríguez-Calcerrada, J., et al., 2019. Breeding and scientific advances in the fight against Dutch elm disease: will they allow the use of elms in forest restoration? *New Forests*, 50(2), pp.183–215. https://doi.org/10.1007/s11056-018-9640-x; Marcone, C., 2017. Elm yellows: a phytoplasma disease of concern in forest and landscape ecosystems. *Forest Pathology*, 47(1), e12324. https://doi.org/10.1111/efp.12324

67. Macaya-Sanz, D., Witzell, J., Collada, C., et al., 2023. Core endophytic mycobiome in *Ulmus minor* and its relation to Dutch elm disease resistance. *Frontiers in Plant Science*, 14, p.1125942. https://doi.org/10.3389/fpls.2023.1125942

68. Kim, H.Y., 2015. Statistical notes for clinical researchers: type I and type II errors in statistical decision. *Restorative Dentistry & Endodontics*, 40(3), pp.249–252. https://doi.org/10.5395/rde.2015.40.3.249

69. Levia, D.F., Nanko, K., Amasaki, H., et al., 2019. Throughfall partitioning by trees. *Hydrological Processes*, 33(12), pp.1698–1708. https://doi.org/10.1002/hyp.13432; Goebes, P., Bruelheide, H., Härdtle, W., et al., 2015. Species-specific effects on throughfall kinetic energy in subtropical forest plantations are related

to leaf traits and tree architecture. *PLoS One*, 10(6), e0128084. https://doi.org/10.1371/journal.pone.0128084

70. Mukherjee, P., Roy, S., Ghosh, D., et al., 2022. Role of animal models in biomedical research: a review. *Laboratory Animals Research*, 38(1), p.18. https://doi.org/10.1186/s42826-022-00128-1

71. Glatigny, S. and Bettelli, E., 2018. Experimental Autoimmune Encephalomyelitis (EAE) as animal models of multiple sclerosis (MS). *Cold Spring Harbor Perspectives in Medicine*, 8(11), a028977. https://doi.org/10.1101/cshperspect.a028977

72. Kappos, L., Bates, D., Edan, G., et al., 2011. Natalizumab treatment for multiple sclerosis: updated recommendations for patient selection and monitoring. *Lancet Neurology*, 10(8), pp.745–758. https://doi.org/10.1016/S1474-4422(11)70149-1

73. Sasaguri, H., Nilsson, P., Hashimoto, S., et al., 2017. APP mouse models for Alzheimer's disease preclinical studies. *The EMBO Journal*, 36(17), pp.2473–2487. https://doi.org/10.15252/embj.201797397

74. Lemon, R.N., 2008. Descending pathways in motor control. *Annual Review of Neuroscience*, 31(1), pp.195–218. https://doi.org/10.1146/annurev.neuro.31.060407.125547

75. Hanson, H.E., Mathews, N.S., Hauber, M.E., et al., 2020. The house sparrow in the service of basic and applied biology. *eLife*, 9, p.e52803. https://doi.org/10.7554/eLife.52803

76. Johnston, R.F. and Selander, R.K., 1964. House sparrows: rapid evolution of races in North America. *Science*, 144(3618), pp.548–550. https://doi.org/10.1126/science.144.3618.548

77. Shefner, J.M., Al-Chalabi, A., Baker, M.R., et al., 2020. A proposal for new diagnostic criteria for ALS. *Clinical Neurophysiology*, 131(8), pp.1975–1978. https://doi.org/10.1016/j.clinph.2020.04.005

78. Bashford, J., Mills, K. and Shaw, C., 2020. The evolving role of surface electromyography in amyotrophic lateral sclerosis: a systematic review. *Clinical Neurophysiology*, 131(4), pp.942–950. https://doi.org/10.1016/j.clinph.2019.12.007

79. Bashford, J., Wickham, A., Iniesta, R., et al., 2019. SPiQE: an automated analytical tool for detecting and characterising fasciculations in amyotrophic lateral sclerosis. *Clinical Neurophysiology*, 130(7), pp.1083–1090. https://doi.org/10.1016/j.clinph.2019.03.032

80. Bashford, J.A., Wickham, A., Iniesta, R., et al., 2020. The rise and fall of fasciculations in amyotrophic lateral sclerosis. *Brain Communications*, 2(1), p.fcaa018. https://doi.org/10.1093/braincomms/fcaa018

81. Ibid.

82. Brabban, A. and Turkington, D., 2002. A casebook of cognitive therapy for psychosis. 1st ed. In: A.P.E. Morrison, ed. *Routledge*.

83. Martín, J.A., Sobrino-Plata, J., Rodríguez-Calcerrada, J., et al., 2019. Breeding and scientific advances in the fight against Dutch elm disease: will they allow the use of elms in forest restoration? *New Forests*, 50(2), pp.183-215. https://doi.org/10.1007/s11056-018-9640-x; Macaya-Sanz, D., Witzell, J., Collada, C., et al., 2023. Core endophytic mycobiome in Ulmus minor and its relation to Dutch elm disease resistance. *Frontiers in Plant Science*, 14, p.1125942. https://doi.org/10.3389/fpls.2023.1125942

84. Martín, J.A. et al., 2019. Breeding and scientific advances in the fight against Dutch elm disease. https://doi.org/10.1007/s11056-018-9640-x

85. Krarup, C., Boeckstyns, M., Ibsen, A., et al., 2016. Remodeling of motor units after nerve regeneration studied by quantitative electromyography. *Clinical Neurophysiology*, 127(2), pp.1675–1682. https://doi.org/10.1016/j.clinph.2015.08.008

86. Stuart-Smith, S., 2020. *The Well Gardened Mind: Rediscovering Nature in the Modern World*. London: HarperCollins Publishers.

Chapter 3

87. World Health Organization (WHO), 2020. Weekly update on COVID-19: 8–15 April 2020. Available at: https://www.who.int/publications/m/item/weekly-update-on-covid-19---15-april-2020

88. Engberg, M., Bonde, J., Sigurdsson, S.T., et al., 2021. Training non-intensivist doctors to work with COVID-19 patients in intensive care units. *Acta Anaesthesiologica Scandinavica*, 65(5), pp.664–673. https://doi.org/10.1111/aas.13789

89. Polack, F.P., Thomas, S.J., Kitchin, N., et al., 2020. Safety and efficacy of the BNT162b2 mRNA Covid-19 vaccine. *New England Journal of Medicine*, 383(27), pp.2603–2615. https://doi.org/10.1056/NEJMoa2034577

90. Hajjar, L.A., Costa, I., Rizk, S.I., et al., 2021. Intensive care management of patients with COVID-19: a practical approach. *Annals of Intensive Care*, 11(1), p.36. https://doi.org/10.1186/s13613-021-00820-w

91. Merriam-Webster, 2020. The history of 'disease'. Available at: https://www.merriam-webster.com/words-at-play/word-history-of-disease

92. UK Government, 2012. Notifiable disease: historic annual totals. Available at: https://www.gov.uk/government/publications/notifiable-diseases-historic-annual-totals; Sharma, N.C., Efstratiou, A., Mokrousov, I., et al., 2019. Diphtheria. *Nature Reviews Disease Primers*, 5(1), p.81. https://doi.org/10.1038/s41572-019-0131-y; Gutierrez, M.C., Brisse, S., Brosch, R., et al., 2005. Ancient origin and gene mosaicism of the progenitor of *Mycobacterium tuberculosis*. *PLoS Pathogens*, 1(1), p.e5. https://doi.org/10.1371/journal.ppat.0010005; Strassburg, M.A., 1982. The global eradication of smallpox. *American Journal of Infection Control*, 10(2), pp.53–59. https://doi.org/10.1016/0196-6553(82)90003-7; Rachlin, A., Patel, J.C., Burns, C.C., et al., 2022. Progress toward polio eradication – worldwide, January 2020–April 2022. *MMWR Morbidity and Mortality Weekly Report*, 71(19), pp.650–655. https://doi.org/10.15585/mmwr.mm7119a2

93. Ritchie, H., 2019. The world population is changing: for the first time there are more people over 64 than children younger than 5. *OurWorldInData. org*. Available at: https://ourworldindata.org/population-aged-65-outnumber-children; World Health Organization (WHO), 2022. Ageing and health. Available at: https://www.who.int/news-room/fact-sheets/detail/ageing-and-health

94. Levine, H.J., 1997. Rest heart rate and life expectancy. *Journal of the American College of Cardiology*, 30(4), pp.1104–1106. https://doi.org/10.1016/s0735-1097(97)00246-5

95. National Institute on Aging (NIA), 2022. Fiscal year 2022 budget – National Institute on Aging. Available at: https://www.nia.nih.gov/about/budget/fiscal-year-2022-budget#graphs

96. Chimombo, S., 1988. The chameleon in lore, life and literature – the poetry of Jack Mapanje. *The Journal of Commonwealth Literature*, 23(1), pp.102–115.; Mapanje, J., 1991. *Of Chameleons and Gods: Poems*. London: Pearson Education.

97. Teyssier, J., Saenko, S.V., van der Marel, D., et al., 2015. Photonic crystals cause active colour change in chameleons. *Nature Communications*, 6, p.6368. https://doi.org/10.1038/ncomms7368

98. Ligon, R.A. and McGraw, K.J., 2013. Chameleons communicate with complex colour changes during contests: different body regions convey different information. *Biology Letters*, 9(6), p.20130892. https://doi.org/10.1098/rsbl.2013.0892; Stuart-Fox, D. and Moussalli, A., 2008. Selection for social signalling drives the evolution of chameleon colour change. *PLoS Biology*, 6(1), p.e25. https://doi.org/10.1371/journal.pbio.0060025; Stuart-Fox, D. and Moussalli, A., 2009. Camouflage, communication and thermoregulation: lessons from colour changing organisms. *Philosophical Transactions of the Royal Society B: Biological Sciences*, 364(1516), pp.463–470. https://doi.org/10.1098/rstb.2008.0254

99. Dahm, R., 2005. Friedrich Miescher and the discovery of DNA. *Developmental Biology*, 278(2), pp.274–288. https://doi.org/10.1016/j.ydbio.2004.11.028; Ghannam, J.Y., Wang, J. and Jan, A., 2023. Biochemistry, DNA Structure. *StatPearls*. Treasure Island, FL: StatPearls Publishing LLC.

100. Callahan, M.P., Smith, K.E., Cleaves, H.J., et al., 2011. Carbonaceous meteorites contain a wide range of extraterrestrial nucleobases. *Proceedings of the National Academy of Sciences*, 108(34), pp.13995–13998. https://doi.org/10.1073/pnas.1106493108

101. Santana-Sagredo, F., Schulting, R.J., Méndez-Quiros, P., et al., 2021. 'White gold' guano fertilizer drove agricultural intensification in the Atacama Desert from AD 1000. *Nature Plants*, 7(2), pp.152–158. https://doi.org/10.1038/s41477-020-00835-4; Bridge, P.J., 1974. Guanine and uricite, two new organic minerals from Peru and Western Australia. *Mineralogical Magazine*, 39(308), pp.889–890.

102. Matos, L.C., Machado, J.P., Monteiro, F.J., et al., 2021. Understanding traditional Chinese medicine therapeutics: an overview of the basics and clinical applications. *Healthcare (Basel)*, 9(3), p.257. https://doi.org/10.3390/healthcare9030257

103. Kolb, S.J. and Kissel, J.T., 2011. Spinal muscular atrophy: a timely review. *Archives of Neurology*, 68(8), pp.979–984. https://doi.org/10.1001/archneurol.2011.74

104. Chaudhry, R. and Khaddour, K., 2023. Biochemistry, DNA Replication. *StatPearls*. Treasure Island, FL: StatPearls Publishing LLC.

105. Durland, J. and Ahmadian-Moghadam, H., 2023. Genetics, Mutagenesis. *StatPearls*. Treasure Island, FL: StatPearls Publishing LLC.

106. Lefebvre, S., Bürglen, L., Reboullet, S., et al., 1995. Identification and characterization of a spinal muscular atrophy-determining gene. *Cell*, 80(1), pp.155–165. https://doi.org/10.1016/0092-8674(95)90460-3

107. Khan, Y.S. and Farhana, A., 2023. Histology, Cell. *StatPearls*. Treasure Island, FL: StatPearls Publishing LLC.

108. Hoerter, J.E. and Ellis, S.R., 2023. Biochemistry, Protein Synthesis. *StatPearls*. Treasure Island, FL: StatPearls Publishing LLC.; Hsieh, M.L. and Borger, J., 2023. Biochemistry, RNA Polymerase. *StatPearls*. Treasure Island, FL: StatPearls Publishing LLC.

109. Chaytow, H., Huang, Y.T., Gillingwater, T.H., et al., 2018. The role of survival motor neuron protein (SMN) in protein homeostasis. *Cellular and Molecular Life Sciences*, 75(21), pp.3877–3894. https://doi.org/10.1007/s00018-018-2849-1

110. Rochette, C., Gilbert, N. and Simard, L., 2001. SMN gene duplication and the emergence of the SMN2 gene occurred in distinct hominids: SMN2 is unique to *Homo sapiens*. *Human Genetics*, 108(3), pp.255–266. https://doi.org/10.1007/s004390100473

111. Burr, P. and Reddivari, A.K.R., 2023. Spinal Muscular Atrophy. *StatPearls*. Treasure Island, FL: StatPearls Publishing LLC.

112. Finkel, R.S., Mercuri, E., Darras, B.T., et al., 2017. Nusinersen versus sham control in infantile-onset spinal muscular atrophy. *New England Journal of Medicine*, 377(18), pp.1723–1732. https://doi.org/10.1056/NEJMoa1702752

113. Ibid.

114. General Medical Council (GMC), 2013. Good Medical Practice. Available at: https://www.gmc-uk.org/ethical-guidance/ethical-guidance-for-doctors/good-medical-practice

115. Tawil, S.E. and Muir, K.W., 2017. Thrombolysis and thrombectomy for acute ischaemic stroke. *Clinical Medicine (London)*, 17(2), pp.161–165. https://doi.org/10.7861/clinmedicine.17-2-161

116. Christos, Y., 2009. Hippocrates of Kos, the Father of clinical medicine, and Asclepiades of Bithynia, the Father of molecular medicine. *In Vivo*, 23(4), pp.507–512.; Hippocrates, 400 BC. *Of the Epidemics (translated by Francis Adams)*. Available at: https://classics.mit.edu/Hippocrates/epidemics.1.i.html

117. Homer, 2020. *The Iliad: A New Translation by Peter Green*. ProQuest Ebook Central. Available at: https://ebookcentral.proquest.com/lib/kcl/detail.action?docID=1882108; Asclepius—man or myth, 2020. *JAMA*, 323(3), p.285. https://doi.org/10.1001/jama.2019.13268

118. Hart, G.D., 1965. Asclepius, God of medicine. *Canadian Medical Association Journal*, 92(5), pp.232–236.

119. Mironidou-Tzouveleki, M. and Tzitzis, P.M., 2014. Medical practice in the ancient Asclepeion in Kos island. *Hellenic Journal of Nuclear Medicine*, 17, pp.167–170.

120. Sacks, A.C. and Michels, R., 2012. Caduceus and Asclepius: history of an error. *American Journal of Psychiatry*, 169(5), p.464. https://doi.org/10.1176/appi.ajp.2012.11121800

121. Orfanos, C.E., 2007. From Hippocrates to modern medicine. *Journal of the European Academy of Dermatology and Venereology*, 21(6), pp.852–858. https://doi.org/10.1111/j.1468-3083.2007.02273.x

122. Christos, Y., 2009. Hippocrates of Kos, the Father of clinical medicine, and Asclepiades of Bithynia, the Father of molecular medicine. *In Vivo*, 23(4), pp.507–512.

123. Santacroce, L., Bottalico, L. and Charitos, I.A., 2017. Greek medicine practice at ancient Rome: The physician molecularist Asclepiades. *Medicines (Basel)*, 4(4), p.92. https://doi.org/10.3390/medicines4040092

124. Prioreschi, P., 1996. *A History of Medicine: Roman Medicine*. Omaha, NE: Horatius Press.

125. Ibid.

126. Ibid.

127. Greive, J., 2010. *A Cornelius Celsus of Medicine*. Read Books Design.

128. Boyd, B., 2018. The evolution of stories: from mimesis to language, from fact to fiction. *Wiley Interdisciplinary Reviews: Cognitive Science*, 9(1), e1444. https://doi.org/10.1002/wcs.1444

129. Harvey, P.D., 2019. Domains of cognition and their assessment. *Dialogues in Clinical Neuroscience*, 21(3), pp.227–237. https://doi.org/10.31887/DCNS.2019.21.3/pharvey

130. McGilchrist, I., 2019. *The Master and His Emissary: The Divided Brain and the Making of the Western World*. New Haven, CT: Yale University Press.

Chapter 4

131. Rehman, I., Gulani, A., Farooq, M., et al., 2023. Genetics, Mitosis. *StatPearls*. Treasure Island, FL: StatPearls Publishing LLC.; Gottlieb, S.F., Gulani, A. and Tegay, D.H., 2023. Genetics, Meiosis. *StatPearls*. Treasure Island, FL: StatPearls Publishing LLC.

132. Bernhardt, H.S., 2012. The RNA world hypothesis: the worst theory of the early evolution of life (except for all the others)(a). *Biology Direct*, 7, p.23. https://doi.org/10.1186/1745-6150-7-23

133. Ghannam, J.Y., Wang, J. and Jan, A., 2023. Biochemistry, DNA Structure. *StatPearls*. Treasure Island, FL: StatPearls Publishing LLC.; Wang, D. and Farhana, A., 2023. Biochemistry, RNA Structure. *StatPearls*. Treasure Island, FL: StatPearls Publishing LLC.

134. Watson, J.D. and Crick, F.H., 1953. Molecular structure of nucleic acids; a structure for deoxyribose nucleic acid. *Nature*, 171(4356), pp.737–738. https://doi.org/10.1038/171737a0

135. Piovesan, A., Pelleri, M.C., Antonaros, F., et al., 2019. On the length, weight and GC content of the human genome. *BMC Research Notes*, 12(1), p.106. https://doi.org/10.1186/s13104-019-4137-z

136. Powner, M.W., Gerland, B. and Sutherland, J.D., 2009. Synthesis of activated pyrimidine ribonucleotides in prebiotically plausible conditions. *Nature*, 459(7244), pp.239–242. https://doi.org/10.1038/nature08013; Benner, S.A., Kim, H.J. and Carrigan, M.A., 2012. Asphalt, water, and the prebiotic synthesis of ribose, ribonucleosides, and RNA. *Accounts of Chemical Research*, 45(12), pp.2025–2034. https://doi.org/10.1021/ar200332w; Vlassov, A.V., Kazakov, S.A., Johnston, B.H., et al., 2005. The RNA world on ice: a new scenario for the emergence of RNA information. *Journal of Molecular Evolution*, 61(2), pp.264–273. https://doi.org/10.1007/s00239-004-0362-7

137. Bernhardt, H.S., 2012. The RNA world hypothesis: the worst theory of the early evolution of life (except for all the others)(a). *Biology Direct*, 7, p.23. https://doi.org/10.1186/1745-6150-7-23

138. Lilley, D.M., 2011. Mechanisms of RNA catalysis. *Philosophical Transactions of the Royal Society B: Biological Sciences*, 366(1580), pp.2910–2917. https://doi.org/10.1098/rstb.2011.0132

139. Ryan, M., Fanning, A., Gerhardt, M.B., et al., 2014. Major winter weather events during the 2013-2014 cold season. Available at: https://www.wpc.ncep.noaa.gov/storm_summaries/event_reviews/2013-2014ColdSeasonArticle.pdf

140. Campbell-Staton, S.C., Cheviron, Z.A., Rochette, N., et al., 2017. Winter storms drive rapid phenotypic, regulatory, and genomic shifts in the green anole lizard. *Science*, 357(6350), pp.495–498. https://doi.org/10.1126/science.aam5512

141. Ibid.

142. Hoerter, J.E. and Ellis, S.R., 2023. Biochemistry, Protein Synthesis. *StatPearls*. Treasure Island, FL: StatPearls Publishing LLC.

143. Quina, A.S., Buschbeck, M. and Di Croce, L., 2006. Chromatin structure and epigenetics. *Biochemical Pharmacology*, 72(11), pp.1563–1569. https://doi.org/10.1016/j.bcp.2006.06.016

144. Ilango, S., Paital, B., Jayachandran, P., et al., 2020. Epigenetic alterations in cancer. *Frontiers in Bioscience*, 25, pp.1058–109. https://doi.org/10.2741/4847

145. Campbell-Staton, S.C., Cheviron, Z.A., Rochette, N., et al., 2017. Winter storms drive rapid phenotypic, regulatory, and genomic shifts in the green anole lizard. *Science*, 357(6350), pp.495–498. https://doi.org/10.1126/science.aam5512

146. Ghosh, A., 2021. *The Nutmeg's Curse: Parables for a Planet in Crisis*. London: John Murray (Publishers).

147. Norton, H.L., Kittles, R.A., Parra, E., et al., 2007. Genetic evidence for the convergent evolution of light skin in Europeans and East Asians. *Molecular Biology and Evolution*, 24(3), pp.710–722. https://doi.org/10.1093/molbev/msl203; Olalde, I., Allentoft, M.E., Sánchez-Quinto, F., et al., 2014. Derived immune and ancestral pigmentation alleles in a 7,000-year-old Mesolithic European. *Nature*, 507(7491), pp.225–228. https://doi.org/10.1038/nature12960; Iain, M., Iosif, L., Nadin, R., et al., 2015. Eight thousand years of natural selection in Europe. *bioRxiv*. https://doi.org/10.1101/016477

148. Jablonski, N.G. and Chaplin, G., 2000. The evolution of human skin coloration. *Journal of Human Evolution*, 39(1), pp.57–106. https://doi.org/10.1006/jhev.2000.0403

149. Chauhan, K., Shahrokhi, M. and Huecker, M.R., 2023. Vitamin D. *StatPearls*. Treasure Island, FL: StatPearls Publishing LLC.

150. Chaplin, G. and Jablonski, N.G., 2009. Vitamin D and the evolution of human depigmentation. *American Journal of Physical Anthropology*, 139(4), pp.451–461. https://doi.org/10.1002/ajpa.21079

151. Gabros, S., Nessel, T.A. and Zito, P.M., 2023. Sunscreens and Photoprotection. *StatPearls*. Treasure Island, FL: StatPearls Publishing LLC.

152. Lucock, M.D., 2023. The evolution of human skin pigmentation: a changing medley of vitamins, genetic variability, and UV radiation during human

expansion. *American Journal of Biological Anthropology*, 180(2), pp.252–271. https://doi.org/10.1002/ajpa.24564

153. Burke, K.D., Williams, J.W., Chandler, M.A., et al., 2018. Pliocene and Eocene provide best analogs for near-future climates. *Proceedings of the National Academy of Sciences*, 115(52), pp.13288–13293. https://doi.org/10.1073/pnas.1809600115; Jablonski, N.G., 2021. The evolution of human skin pigmentation involved the interactions of genetic, environmental, and cultural variables. *Pigment Cell & Melanoma Research*, 34(4), pp.707–729. https://doi.org/10.1111/pcmr.12976

154. Dror, Y. and Hopp, M., 2014. Hair for brain trade-off, a metabolic bypass for encephalization. *SpringerPlus*, 3, p.562. https://doi.org/10.1186/2193-1801-3-562

155. Sandel, A.A., 2013. Brief communication: Hair density and body mass in mammals and the evolution of human hairlessness. *American Journal of Physical Anthropology*, 152(1), pp.145–150. https://doi.org/10.1002/ajpa.22333

156. Lopez, M.J. and Mohiuddin, S.S., 2023. Biochemistry, Essential Amino Acids. *StatPearls*. Treasure Island, FL: StatPearls Publishing LLC.

157. Dror, Y. and Hopp, M., 2014. Hair for brain trade-off, a metabolic bypass for encephalization. *SpringerPlus*, 3, p.562. https://doi.org/10.1186/2193-1801-3-562

158. Morris, A. and Patel, G., 2023. Heat Stroke. *StatPearls*. Treasure Island, StatPearls Publishing LLC.

159. Best, A. and Kamilar, J.M., 2018. The evolution of eccrine sweat glands in human and nonhuman primates. *Journal of Human Evolution*, 117, pp.33–43. https://doi.org/10.1016/j.jhevol.2017.12.003

160. D'Alba, L. and Shawkey, M.D., 2019. Melanosomes: biogenesis, properties, and evolution of an ancient organelle. *Physiological Reviews*, 99(1), pp.1–19. https://doi.org/10.1152/physrev.00059.2017

161. Solano, F., 2014. Melanins: skin pigments and much more – Types, structural models, biological functions, and formation routes. *New Journal of Science*, 2014, p.498276. https://doi.org/10.1155/2014/498276

162. Gabros, S., Nessel, T.A. and Zito, P.M., 2023. Sunscreens and Photoprotection. *StatPearls*. Treasure Island, FL: StatPearls Publishing LLC.

163. Merrell, B.J. and McMurry, J.P., 2023. Folic Acid. *StatPearls*. Treasure Island, FL: StatPearls Publishing LLC.

164. Lucock, M.D., 2023. The evolution of human skin pigmentation: a changing medley of vitamins, genetic variability, and UV radiation during human expansion. *American Journal of Biological Anthropology*, 180(2), pp.252–271. https://doi.org/10.1002/ajpa.24564

165. Jablonski, N.G. and Chaplin, G., 2000. The evolution of human skin coloration. *Journal of Human Evolution*, 39(1), pp.57–106. https://doi.org/10.1006/jhev.2000.0403

166. Chaplin, G. and Jablonski, N.G., 2009. Vitamin D and the evolution of human depigmentation. *American Journal of Physical Anthropology*, 139(4), pp.451–461. https://doi.org/10.1002/ajpa.21079

167. Norton, H.L., Kittles, R.A., Parra, E., et al., 2007. Genetic evidence for the convergent evolution of light skin in Europeans and East Asians. *Molecular Biology and Evolution*, 24(3), pp.710–722. https://doi.org/10.1093/molbev/msl203; Iain, M., Iosif, L., Nadin, R., et al., 2015. Eight thousand years of natural selection in Europe. *bioRxiv*. https://doi.org/10.1101/016477; Rocha, J., 2020. The evolutionary history of human skin pigmentation. *Journal of Molecular Evolution*, 88(1), pp.77–87. https://doi.org/10.1007/s00239-019-09902-7

168. Lambeck, K., Rouby, H., Purcell, A., et al., 2014. Sea level and global ice volumes from the Last Glacial Maximum to the Holocene. *Proceedings of the National Academy of Sciences*, 111(43), pp.15296–15303. https://doi.org/10.1073/pnas.1411762111

169. Tresset, A. and Vigne, J.D., 2011. Last hunter-gatherers and first farmers of Europe. *Compte Rendus Biologies*, 334(3), pp.182–189. https://doi.org/10.1016/j.crvi.2010.12.010

170. Glisson, F., Bate, G. and Regemorter, A., 1651. *A treatise of the rickets: being a disease common to children*. Translated into English by Phil. Armin. Available at: https://wellcomecollection.org/works/qn24bpgf

171. Chesney, R.W., 2012. Environmental factors in Tiny Tim's near-fatal illness. *Archives of Pediatrics & Adolescent Medicine*, 166(3), pp.271–275. https://doi.org/10.1001/archpediatrics.2011.852; Dahash, B.A. and Sankararaman, S., 2023. Rickets. *StatPearls*. Treasure Island, FL: StatPearls Publishing LLC.

172. Chauhan, K., Shahrokhi, M. and Huecker, M.R., 2023. Vitamin D. *StatPearls*. Treasure Island, FL: StatPearls Publishing LLC.

173. Arnsten, A., Mazure, C.M. and Sinha, R., 2012. This is your brain in meltdown. *Scientific American*, 306(4), pp.48–53.

174. Libretti, S. and Puckett, Y., 2023. Physiology, Homeostasis. *StatPearls*. Treasure Island, FL: StatPearls Publishing LLC.

175. Osilla, E.V., Marsidi, J.L., Shumway, K.R., et al., 2023. Physiology, Temperature Regulation. *StatPearls*. Treasure Island, FL: StatPearls Publishing LLC.

176. Balli, S., Shumway, K.R. and Sharan, S., 2023. Physiology, Fever. *StatPearls*. Treasure Island, FL: StatPearls Publishing LLC.

177. Wrotek, S., LeGrand, E.K., Dzialuk, A., et al., 2021. Let fever do its job: the meaning of fever in the pandemic era. *Evolutionary Medicine and Public Health*, 9(1), pp.26–35. https://doi.org/10.1093/emph/eoaa044

178. Rosenberg, H., Pollock, N., Schiemann, A., et al., 2015. Malignant hyperthermia: a review. *Orphanet Journal of Rare Diseases*, 10, p.93. https://doi.org/10.1186/s13023-015-0310-1

179. Pham, S. and Puckett, Y., 2023. Physiology, Skeletal Muscle Contraction. *StatPearls*. Treasure Island, FL: StatPearls Publishing LLC.

180. Rosenberg, H., Pollock, N., Schiemann, A., et al., 2015. Malignant hyperthermia: a review. *Orphanet Journal of Rare Diseases*, 10, p.93. https://doi.org/10.1186/s13023-015-0310-1

181. Christos, Y., 2009. Hippocrates of Kos, the Father of clinical medicine, and Asclepiades of Bithynia, the Father of molecular medicine. *In Vivo*, 23(4), pp.507–512.

182. Intergovernmental Panel on Climate Change, 2023. Climate change synthesis report. Available at: https://doi.org/10.59327/IPCC/AR6-9789291691647

183. Cowie, R.H., Bouchet, P. and Fontaine, B., 2022. The Sixth Mass Extinction: fact, fiction or speculation? *Biological Reviews of the Cambridge Philosophical Society*, 97(2), pp.640–663. https://doi.org/10.1111/brv.12816

184. Zackrisson, E., Calissendo, P., Gonzalez, J., et al., 2016. Terrestrial planets across space and time. *arXiv*. https://doi.org/10.3847/1538-4357/833/2/214

185. Cowie, R.H., Bouchet, P. and Fontaine, B., 2022. The Sixth Mass Extinction: fact, fiction or speculation? *Biological Reviews of the Cambridge Philosophical Society*, 97(2), pp.640-663. https://doi.org/10.1111/brv.12816

186. Ripple, W.J., Wolf, C., Lenton, T.M., et al., 2023. Many risky feedback loops amplify the need for climate action. *One Earth*, 6(2), pp.86–91. https://doi.org/10.1016/j.oneear.2023.01.004

187. Willcock, S., Cooper, G.S., Addy, J., et al., 2023. Earlier collapse of Anthropocene ecosystems driven by multiple faster and noisier drivers. *Nature Sustainability*, 6, pp.1331–1342. https://doi.org/10.1038/s41893-023-01157-x

188. Rockström, J., 2023. Planetary boundaries: scientific advances. Available at: https://www.youtube.com/watch?v=7KfWGAjJAsM; Rockström, J., 2023. How 16 tipping points could push our entire planet into crisis. *World Economic Forum*. Available at: https://www.weforum.org/videos/how-16-tipping-points-could-push-our-entire-planet-into-crisis/

189. Randers, J. and Goluke, U., 2020. An earth system model shows self-sustained thawing of permafrost even if all man-made GHG emissions stop in 2020. *Scientific Reports*, 10(1), p.18456. https://doi.org/10.1038/s41598-020-75481-z

Chapter 5

190. NHS England, 2013. A&E attendances and emergency admissions. Available at: https://www.england.nhs.uk/statistics/statistical-work-areas/ae-waiting-times-and-activity/weekly-ae-sitreps-2012-13/; Department of Health and Social Care, 2013. Press release. Available at: https://www.gov.uk/government/news/south-london-healthcare-nhs-trust-to-be-dissolved-by-1-october-2013

191. Skryabina, E., Betts, N., Reedy, G., et al., 2021. UK healthcare staff experiences and perceptions of a mass casualty terrorist incident response: a mixed-methods study. *Emergency Medicine Journal*, 38(10), pp.756–764. https://doi.org/10.1136/emermed-2019-208966; Earthquake Hazards Program, n.d. Earthquakes with 50,000 or more deaths. Available at: https://web.archive.org/web/20130507101448/http://earthquake.usgs.gov/earthquakes/world/most_destructive.php

192. Dalmau, J., Tüzün, E., Wu, H.Y., et al., 2007. Paraneoplastic anti-N-methyl-D-aspartate receptor encephalitis associated with ovarian teratoma. *Annals of Neurology*, 61(1), pp.25–36. https://doi.org/10.1002/ana.21050

193. Samanta, D., Lui, F., 2023. Anti-NMDAR Encephalitis. *StatPearls*. Treasure Island, FL: StatPearls Publishing LLC.

194. Vaillant, A.A.J., Sabir, S. and Jan, A., 2022. Physiology, Immune Response. *StatPearls*. Treasure Island, FL: StatPearls Publishing LLC.

195. Aziz, M., Iheanacho, F. and Hashmi, M.F., 2023. Physiology, Antibody. *StatPearls*. Treasure Island, FL: StatPearls Publishing LLC.

196. Uy, C.E., Binks, S. and Irani, S.R., 2021. Autoimmune encephalitis: clinical spectrum and management. *Practical Neurology*, 21(5), pp.412–423. https://doi.org/10.1136/practneurol-2020-002567

197. Dalmau, J., Gleichman, A.J., Hughes, E.G., et al., 2008. Anti-NMDA-receptor encephalitis: case series and analysis of the effects of antibodies. *Lancet Neurology*, 7(12), pp.1091–1098. https://doi.org/10.1016/S1474-4422(08)70224-2; Jewett, B.E. and Thapa, B., 2023. Physiology, NMDA Receptor. *StatPearls*. Treasure Island, FL: StatPearls Publishing LLC.

198. Rockefeller University, 1926. Monographs of the Rockefeller Institute for Medical Research, Vol. 21.; Medawar, P.B., 1948. Immunity to homologous grafted skin; the fate of skin homografts transplanted to the brain, to subcutaneous tissue, and to the anterior chamber of the eye. *British Journal of Experimental Pathology*, 29(1), pp.58–69.

199. Hong, S. and Van Kaer, L., 1999. Immune privilege: keeping an eye on natural killer T cells. *Journal of Experimental Medicine*, 190(9), pp.1197–1200. https://doi.org/10.1084/jem.190.9.1197

200. Filiano, A.J., Gadani, S.P. and Kipnis, J., 2015. Interactions of innate and adaptive immunity in brain development and function. *Brain Research*, 1617, pp.18–27. https://doi.org/10.1016/j.brainres.2014.07.050

201. Mapunda, J.A., Tibar, H., Regragui, W., et al., 2022. How does the immune system enter the brain? *Frontiers in Immunology*, 13, p.805657. https://doi.org/10.3389/fimmu.2022.805657

202. Samanta, D., Lui, F., 2023. Anti-NMDAR Encephalitis. *StatPearls*. Treasure Island, FL: StatPearls Publishing LLC.

203. Joshi, M., 1997. *Vadiraja Theertha Prabandha: An Unique Travel Guide of 107 Holy Kshetras*, p.93.

204. Vanamali, 2002. *Shiva: Stories and Teachings from the Shiva Mahapurana*, p.172.

205. Rogers, J.J.W. and Santosh, M., 2004. *Continents and Supercontinents*. Oxford: ProQuest Ebook Central, Oxford University Press.

206. Hawkesworth, C.J. and Brown, M., 2018. Earth dynamics and the development of plate tectonics. *Philosophical Transactions of the Royal Society A: Mathematical, Physical and Engineering Sciences*, 376(2132). https://doi.org/10.1098/rsta.2018.0228

207. Kumar, P., Yuan, X., Kumar, M.R., et al., 2007. The rapid drift of the Indian tectonic plate. *Nature*, 449(7164), pp.894–897. https://doi.org/10.1038/nature06214

208. Rogers, J.J.W. and Santosh, M., 2004. *Continents and Supercontinents*.

209. Turner, B., Jenkins, J., Turner, R., et al., 2013. *Seismicity of the Earth 1900–2010 Himalaya and vicinity*. Open-File Report, Originally posted 11 March 2013; Revised 28 January 2014, ed. V.A. Reston.

210. Kumar, P., Yuan, X., Kumar, M.R., et al., 2007. The rapid drift of the Indian tectonic plate. *Nature*, 449(7164), pp.894–897. https://doi.org/10.1038/nature06214

211. Pérez-Díaz, L., Eagles, G. and Sigloch, K., 2020. Indo-Atlantic plate accelerations around the Cretaceous-Paleogene boundary: a time-scale error, not a plume-push signal. *Geology*, 48(12), pp.1169–1173. https://doi.org/10.1130/G47859.1

212. Andrews, R.G., 2021. The new historian of the smash that made the Himalayas. *Quanta Magazine*. Available at: https://www.quantamagazine.org/the-new-historian-of-the-smash-that-made-the-himalayas-20210414/

213. Ibid.

214. Pérez-Díaz, L., Eagles, G. and Sigloch, K., 2020. Indo-Atlantic plate accelerations around the Cretaceous-Paleogene boundary: a time-scale error, not a plume-push signal. *Geology*, 48(12), pp.1169–1173. https://doi.org/10.1130/G47859.1

215. Hawkesworth, C.J. and Brown, M., 2018. Earth dynamics and the development of plate tectonics. *Philosophical Transactions of the Royal Society A: Mathematical, Physical and Engineering Sciences*, 376(2132). https://doi.org/10.1098/rsta.2018.0228

216. BBC Media Centre, 2023. Chris Packham on Earth: Q&A. Available at: https://www.bbc.co.uk/mediacentre/mediapacks/earth

217. Birkel, S.D., n.d. Daily sea surface temperature. Climate Reanalyzer: Climate Change Institute, University of Maine, USA. https://climatereanalyzer.org/clim/sst_daily/?dm_id=world2

218. Osman, M.B., Coats, S., Das, S.B., et al., 2021. North Atlantic jet stream projections in the context of the past 1,250 years. *Proceedings of the National Academy of Sciences of the United States of America*, 118(38). https://doi.org/10.1073/pnas.2104105118

219. Woollings, T., Drouard, M., O'Reilly, C.H., et al., 2023. Trends in the atmospheric jet streams are emerging in observations and could be linked to tropical warming. *Communications Earth & Environment*, 4(1), p.125. https://doi.org/10.1038/s43247-023-00792-8

220. Nakamura, N. and Huang, C.S.Y., 2018. Atmospheric blocking as a traffic jam in the jet stream. *Science*, 361(6397), pp.42–47. https://doi.org/10.1126/science.aat0721; Climate Change Service, 2023. July 2023 sees multiple global temperature records broken. *Copernicus*. Available at: https://climate.copernicus.eu/july-2023-sees-multiple-global-temperature-records-broken

221. Clark, P.U., Pisias, N.G., Stocker, T.F., et al., 2002. The role of the thermohaline circulation in abrupt climate change. *Nature*, 415(6874), pp.863–869. https://doi.org/10.1038/415863a

222. Ditlevsen, P. and Ditlevsen, S., 2023. Warning of a forthcoming collapse of the Atlantic meridional overturning circulation. *Nature Communications*, 14(1), p.4254. https://doi.org/10.1038/s41467-023-39810-w; Rahmstorf, S., Box, J.E., Feulner, G.,

et al., 2015. Exceptional twentieth-century slowdown in Atlantic Ocean overturning circulation. *Nature*, 5, pp.475–480. https://doi.org/10.1038/nclimate2554

223. Latif, M., Sun, J., Visbeck, M., et al., 2022. Natural variability has dominated Atlantic Meridional Overturning Circulation since 1900. *Nature Climate Change*, 12(5), pp.455–460. https://doi.org/10.1038/s41558-022-01342-4

224. Caesar, L., Rahmstorf, S., Robinson, A., et al., 2018. Observed fingerprint of a weakening Atlantic Ocean overturning circulation. *Nature*, 556(7700), pp.191–196. https://doi.org/10.1038/s41586-018-0006-5

225. Ditlevsen, P. and Ditlevsen, S., 2023. Warning of a forthcoming collapse of the Atlantic meridional overturning circulation. *Nature Communications*, 14(1), p.4254. https://doi.org/10.1038/s41467-023-39810-w

226. Bashford, J., Chan, W.K., Coutinho, E., et al., 2021. Demystifying the spontaneous phenomena of motor hyperexcitability. *Clinical Neurophysiology*, 132(8), pp.1830–1844. https://doi.org/10.1016/j.clinph.2021.03.053

227. Sacks, J., 2016. *Not in God's Name: Confronting Religious Violence*. New Milford, CT: Maggid Books.

228. Warren, J.D., Hardy, C.J., Fletcher, P.D., et al., 2016. Binary reversals in primary progressive aphasia. *Cortex*, 82, pp.287–289. https://doi.org/10.1016/j.cortex.2016.05.017

229. Sacks, J., 2016. *Not in God's Name: Confronting Religious Violence*. New Milford, CT: Maggid Books.

230. Vanamali, 2002. *Shiva: Stories and Teachings from the Shiva Mahapurana*, p.172.

231. Wiggins, G.A., Tyack, P., Scharff, C., et al., 2015. The evolutionary roots of creativity: mechanisms and motivations. *Philosophical Transactions of the Royal Society B: Biological Sciences*, 370(1664), p.20140099. https://doi.org/10.1098/rstb.2014.0099

232. Zmigrod, L., 2020. The role of cognitive rigidity in political ideologies: Theory, evidence, and future directions. *Current Opinion in Behavioral Sciences*, 34, pp.34–39. https://doi.org/10.1016/j.cobeha.2019.10.016; van Baar, J.M. and Feldman Hall, O., 2022. The polarized mind in context. *Interdisciplinary approaches to the psychology of political polarization*, 77(3), pp.394–408. https://doi.org/10.1037/amp0000814

233. Yale Project on Climate Change, 2009. Global warming's Six Americas 2009: an audience segmentation analysis. Available at: https://climatecommunication.yale.edu/wp-content/uploads/2016/02/2009_05_Global-Warmings-Six-Americas.pdf; Hayhoe, K., 2021. *Saving Us: A Climate Scientist's Case for Hope and Healing in a Divided World*. New York: Atria/One Signal Publishers.

234. Yale School of the Environment, 2022. Global Warming's Six Americas. Yale Program on climate change communication. Available at: https://climatecommunication.yale.edu/about/projects/global-warmings-six-americas/

235. Tsai, J.F., Kudo, S. and Yoshizawa, K., 2015. Maternal care in Acanthosomatinae (Insecta: Heteroptera: Acanthosomatidae)—correlated evolution with morphological change. *BMC Evolutionary Biology*, 15, p.258. https://doi.org/10.1186/s12862-015-0537-4

236. Natural History Museum, n.d. Meet the bee-fly: the cute bee mimic with a dark side. Available at: https://www.nhm.ac.uk/discover/bee-flies-cute-bee-mimic-with-a-dark-side.html

Chapter 6

237. Asnicar, F., Berry, S.E., Valdes, A.M., et al., 2021. Microbiome connections with host metabolism and habitual diet from 1,098 deeply phenotyped individuals. *Nature Medicine*, 27(2), pp.321–332. https://doi.org/10.1038/s41591-020-01183-8

238. Pecoits, E., Konhauser, K.O., Aubet, N.R., et al., 2012. Bilaterian burrows and grazing behavior at >585 million years ago. *Science*, 336(6089), pp.1693–1696. https://doi.org/10.1126/science.1216295

239. Khan, Y.S. and Ackerman, K.M., 2023. Embryology, Week 1. *StatPearls*. Treasure Island, FL: StatPearls Publishing LLC.

240. Maderspacher, F., 2009. Breakthroughs and blind ends. *Current Biology*, 19(7), pp.R272-R274.

241. Sheldrake, M., 2020. *Entangled Life: How Fungi Make Our Worlds, Change Our Minds and Shape Our Futures.* London: Random House.

242. Hejnol, A. and Martín-Durán, J.M., 2015. Getting to the bottom of anal evolution. *Zoologischer Anzeiger – A Journal of Comparative Zoology*, 256, pp.61–74. https://doi.org/10.1016/j.jcz.2015.02.006

243. Arendt, D., 2003. Spiralians in the limelight. *Genome Biology*, 5(1), p.303. https://doi.org/10.1186/gb-2003-5-1-303

244. Nielsen, C., Brunet, T. and Arendt, D., 2018. Evolution of the bilaterian mouth and anus. *Nature Ecology & Evolution*, 2(9), pp.1358–1376. https://doi.org/10.1038/s41559-018-0641-0

245. Heger, P., Zheng, W., Rottmann, A., et al., 2020. The genetic factors of bilaterian evolution. *Elife*, 9, e45530. https://doi.org/10.7554/eLife.45530

246. Genikhovich, G. and Technau, U., 2017. On the evolution of bilaterality. *Development*, 144(19), pp.3392–3404. https://doi.org/10.1242/dev.141507

247. Hejnol, A. and Martin-Duran, J.M. 2015. Getting to the bottom of anal evolution. *Zoologischer Anzeiger - A Journal of Comparative Zoology*, 256, pp.61–74. https://doi.org/10.1016/j.jcz.2015.02.006

248. Nuttall, G.H.F. and Thierfelder, H., 1897. Thierisches Leben ohne Bakterien im Verdauungskanal. (II. Mittheilung). *Archiv für Mikrobiologie*, 22(1), pp.62–73.; Thompson, G.R. and Trexler, P.C., 1971. Gastrointestinal structure and function in germ-free or gnotobiotic animals. *Gut*, 12(3), pp.230–235. https://doi.org/10.1136/gut.12.3.230

249. Luczynski, P., McVey Neufeld, K.A., Oriach, C.S., et al., 2016. Growing up in a bubble: using germ-free animals to assess the influence of the gut microbiota on brain and behavior. *International Journal of Neuropsychopharmacology*, 19(8). https://doi.org/10.1093/ijnp/pyw020

250. Balestrino, R. and Schapira, A.H.V., 2020. Parkinson disease. *European Journal of Neurology*, 27(1), pp.27–42. https://doi.org/10.1111/ene.14108

251. Rees, R.N., Noyce, A.J. and Schrag, A., 2019. The prodromes of Parkinson's disease. *European Journal of Neuroscience*, 49(3), pp.320–327. https://doi.org/10.1111/ejn.14269; Schrag, A., Bohlken, J., Dammertz, L., et al., 2023. Widening the spectrum of risk factors, comorbidities, and prodromal features of Parkinson disease. *JAMA Neurology*, 80(2), pp.161–171. https://doi.org/10.1001/jamaneurol.2022.3902

252. Wallen, Z.D., Demirkan, A., Twa, G., et al., 2022. Metagenomics of Parkinson's disease implicates the gut microbiome in multiple disease mechanisms. *Nature Communications*, 13(1), p.6958. https://doi.org/10.1038/s41467-022-34667-x

253. Romano, S., Savva, G.M., Bedarf, J.R., et al., 2021. Meta-analysis of the Parkinson's disease gut microbiome suggests alterations linked to intestinal inflammation. *NPJ Parkinson's Disease*, 7(1), p.27. https://doi.org/10.1038/s41531-021-00156-z

254. Thursby, E. and Juge, N., 2017. Introduction to the human gut microbiota. *Biochemical Journal*, 474(11), pp.1823–1836. https://doi.org/10.1042/BCJ20160510

255. Lewandowska-Pietruszka, Z., Figlerowicz, M. and Mazur-Melewska, K., 2022. The history of the intestinal microbiota and the gut-brain axis. *Pathogens*, 11(12), p.1540. https://doi.org/10.3390/pathogens11121540; Kenny, B.J. and Bordoni,

B., 2023. Neuroanatomy, Cranial Nerve 10 (Vagus Nerve). *StatPearls*. Treasure Island, FL: StatPearls Publishing LLC.

256. Tan, A.H., Lim, S.Y. and Lang, A.E., 2022. The microbiome–gut–brain axis in Parkinson disease—from basic research to the clinic. *Nature Reviews Neurology*, 18(8), pp.476–495. https://doi.org/10.1038/s41582-022-00681-2; Rietdijk, C.D., Perez-Pardo, P., Garssen, J., et al., 2017. Exploring Braak's hypothesis of Parkinson's disease. *Frontiers in Neurology*, 8, p.37. https://doi.org/10.3389/fneur.2017.00037

257. Plassais, J., Gbikpi-Benissan, G., Figarol, M., et al., 2021. Gut microbiome alpha-diversity is not a marker of Parkinson's disease and multiple sclerosis. *Brain Communications*, 3(2), p.fcab113. https://doi.org/10.1093/braincomms/fcab113; Li, Z., Zhou, J., Liang, H., et al., 2022. Differences in alpha diversity of gut microbiota in neurological diseases. *Frontiers in Neuroscience*, 16, p.879318. https://doi.org/10.3389/fnins.2022.879318

258. Kawahara, A.Y., Plotkin, D., Espeland, M., et al., 2019. Phylogenomics reveals the evolutionary timing and pattern of butterflies and moths. *Proceedings of the National Academy of Sciences*, 116(45), pp.22657–22663. https://doi.org/10.1073/pnas.1907847116

259. Smithsonian Institution, 2021. Moths (BugInfo). Available at: https://www.si.edu/spotlight/buginfo/moths#:~:text=Moths%20are%20in%20the%20insect,to%2017%2C500%20species%20of%20butterflies

260. Sokal, R.R. and Crovello, T.J., 1970. The biological species concept: a critical evaluation. *The American Naturalist*, 104(936), pp.127–153. https://doi.org/10.1086/282646

261. de Queiroz, K., 2005. Ernest Mayr and the modern concept of species. *Proceedings of the National Academy of Sciences*, 102(suppl_1), pp.6600–6607. https://doi.org/10.1073/pnas.0502030102

262. Wake, D.B., Yanev, K.P. and Brown, C.W., 1986. Intraspecific sympatry in a "Ring Species," the plethodontid salamander *Ensatina eschscholtzii*, in Southern California. *Evolution*, 40(4), pp.866–868. https://doi.org/10.1111/j.1558-5646.1986.tb00548.x

263. Moritz, C., Schneider, C.J. and Wake, D.B., 1992. Evolutionary relationships within the *Ensatina Eschscholtzii* complex confirm the ring species interpretation. *Systematic Biology*, 41(3), pp.273–291. https://doi.org/10.1093/sysbio/41.3.273

264. Paterlini, M., 2007. There shall be order. The legacy of Linnaeus in the age of molecular biology. *EMBO Reports*, 8(9), pp.814–816. https://doi.org/10.1038/sj.embor.7401061

265. Sato, A., Tichy, H., O'HUigin, C., et al., 2001. On the origin of Darwin's finches. *Molecular Biology and Evolution*, 18(3), pp.299–311. https://doi.org/10.1093/oxfordjournals.molbev.a003806

266. Roswell, M., Dushoff, J. and Winfree, R., 2021. A conceptual guide to measuring species diversity. *Oikos*, 130(3), pp.321–338. https://doi.org/10.1111/oik.07202

267. Purvis, A. and Hector, A., 2000. Getting the measure of biodiversity. *Nature*, 405(6783), pp.212–219. https://doi.org/10.1038/35012221

268. Schumm, M., Edie, S.M., Collins, K.S., et al., 2019. Common latitudinal gradients in functional richness and functional evenness across marine and terrestrial systems. *Proceedings of the Royal Society B: Biological Sciences*, 286(1908), p.20190745. https://doi.org/10.1098/rspb.2019.0745

269. Leitão, R.P., Zuanon, J., Villéger, S., et al., 2016. Rare species contribute disproportionately to the functional structure of species assemblages. *Proceedings of the Royal Society B: Biological Sciences*, 283(1828), p.20160084. https://doi.org/10.1098/rspb.2016.0084; Wilson, E.O., 1992. *The Diversity of Life*. Cambridge, MA: Belknap Press of Harvard University Press.

270. Hagen, O., Flück, B., Fopp, F., et al., 2021. gen3sis: a general engine for eco-evolutionary simulations of the processes that shape Earth's biodiversity. *PLoS Biology*, 19(7), p.e3001340. https://doi.org/10.1371/journal.pbio.3001340

271. Wagg, C., Roscher, C., Weigelt, A., et al., 2022. Biodiversity–stability relationships strengthen over time in a long-term grassland experiment. *Nature Communications*, 13(1), p.7752. https://doi.org/10.1038/s41467-022-35189-2

272. Weisser, W.W., Roscher, C., Meyer, S.T., et al., 2017. Biodiversity effects on ecosystem functioning in a 15-year grassland experiment: patterns, mechanisms, and open questions. *Basic and Applied Ecology*, 23, pp.1–73. https://doi.org/10.1016/j.baae.2017.06.002

273. Convention on Biological Diversity, 2022. Kunming-Montreal Global Biodiversity Framework. Available at: https://www.cbd.int/gbf/

274. State of Nature Partnership, 2023. State of Nature report. Available at: https://www.stateofnature.org.uk

275. Sutherland, N. and Ares, E., 2022. Protecting and restoring nature at CoP15 and beyond. *House of Commons Library*. Available at: https://commonslibrary.parliament.uk/research-briefings/cdp-2022-0056/; Scholes, R.J. and Biggs, R., 2005. A biodiversity intactness index. *Nature*, 434(7029), pp.45–49. https://doi.org/10.1038/nature03289; Joint Nature Conservation Committee, 2022. UK biodiversity indicators. Available at: https://jncc.gov.uk/our-work/uk-biodiversity-indicators

276. State of Nature Partnership, 2023. State of Nature report. Available at: https://www.stateofnature.org.uk

277. Macfarlane, R. and Morris, J., 2017. *The Lost Words: A Spell Book*. London: Hamish Hamilton.

Chapter 7

278. Capaldi, C.A., Dopko, R.L. and Zelenski, J.M., 2014. The relationship between nature connectedness and happiness: a meta-analysis. *Frontiers in Psychology*, 5, p.976. https://doi.org/10.3389/fpsyg.2014.00976; Schultz, P.W. and Tabanico, J., 2007. Self, identity, and the natural environment: Exploring implicit connections with nature. *Journal of Applied Social Psychology*, 37(6), pp.1219–1247. https://doi.org/10.1111/j.1559-1816.2007.00210.x

279. Nisbet, E.K. and Zelenski, J.M., 2013. The NR-6: a new brief measure of nature relatedness. *Frontiers in Psychology*, 4, p.813. https://doi.org/10.3389/fpsyg.2013.00813

280. Capaldi, C.A., Dopko, R.L. and Zelenski, J.M., 2014. The relationship between nature connectedness and happiness: a meta-analysis. *Frontiers in Psychology*, 5, p.976. https://doi.org/10.3389/fpsyg.2014.00976; Soga, M. and Gaston, K.J., 2023. Global synthesis reveals heterogeneous changes in connection of humans to nature. *One Earth*, 6(2), pp.131–138. https://doi.org/10.1016/j.oneear.2023.01.007

281. Wilson, E.O., 1984. *Biophilia*. Cambridge, MA: Harvard University Press.

282. Chang, C.C., Cox, D.T.C., Fan, Q., et al., 2022. People's desire to be in nature and how they experience it are partially heritable. *PLoS Biology*, 20(2), p.e3001500. https://doi.org/10.1371/journal.pbio.3001500

283. Sahu, M. and Prasuna, J.G., 2016. Twin studies: a unique epidemiological tool. *Indian Journal of Community Medicine*, 41(3), pp.177–182. https://doi.org/10.4103/0970-0218.183593

284. Chang, C.C., Cox, D.T.C., Fan, Q., et al., 2022. People's desire to be in nature and how they experience it are partially heritable. *PLoS Biology*, 20(2), p.e3001500. https://doi.org/10.1371/journal.pbio.3001500

285. Twohig-Bennett, C. and Jones, A., 2018. The health benefits of the great outdoors: a systematic review and meta-analysis of greenspace exposure and health outcomes. *Environmental Research*, 166, pp.628–637. https://doi.org/10.1016/j. envres.2018.06.030

286. Spehar, B., Clifford, C.W.G., Newell, B.R., et al., 2003. Universal aesthetic of fractals. *Computers & Graphics*, 27(5), pp.813–820. https://doi.org/10.1016/ S0097-8493(03)00154-7

287. Taylor, R., Spehar, B., Hagerhall, C., et al., 2011. Perceptual and physiological responses to Jackson Pollock's fractals. *Frontiers in Human Neuroscience*, 5, p.60. https://doi.org/10.3389/fnhum.2011.00060; Taylor, R.P., 2021. The potential of biophilic fractal designs to promote health and performance: a review of experiments and applications. *Sustainability*, 13(2), p.823. https://doi. org/10.3390/su13020823; Hagerhall, C.M., Laike, T., Taylor, R.P., et al., 2008. Investigations of human EEG response to viewing fractal patterns. *Perception*, 37(10), pp.1488–1494. https://doi.org/10.1068/p5918

288. Bales, T.R., Lopez, M.J. and Clark, J., 2023. Embryology, Eye. *StatPearls*. Treasure Island, FL: StatPearls Publishing LLC.

289. Motlagh, M. and Geetha, R., 2023. Physiology, Accommodation. *StatPearls*. Treasure Island, FL: StatPearls Publishing LLC.

290. Ott, M., 2006. Visual accommodation in vertebrates: mechanisms, physiological response and stimuli. *Journal of Comparative Physiology A*, 192(2), pp.97–111. https://doi.org/10.1007/s00359-005-0049-6

291. Holden, B.A., Fricke, T.R., Wilson, D.A., et al., 2016. Global prevalence of myopia and high myopia and temporal trends from 2000 through 2050. *Ophthalmology*, 123(5), pp.1036–1042. https://doi.org/10.1016/j.ophtha.2016.01.006

292. Lingham, G., Mackey, D.A., Lucas, R., et al., 2020. How does spending time outdoors protect against myopia? A review. *British Journal of Ophthalmology*, 104(5), pp.593–599. https://doi.org/10.1136/bjophthalmol-2019-314675

293. Tkatchenko, T.V., Troilo, D., Benavente-Perez, A., et al., 2018. Gene expression in response to optical defocus of opposite signs reveals bidirectional mechanism of visually guided eye growth. *PLoS Biology*, 16(10), p.e2006021. https://doi. org/10.1371/journal.pbio.2006021

294. Kiefer, A.K., Tung, J.Y., Do, C.B., et al., 2013. Genome-wide analysis points to roles for extracellular matrix remodeling, the visual cycle, and neuronal development in myopia. *PLoS Genetics*, 9(2), p.e1003299. https://doi.org/10.1371/journal. pgen.1003299; Zhang, X., Qu, X. and Zhou, X., 2015. Association between parental myopia and the risk of myopia in a child. *Experimental and Therapeutic Medicine*, 9(6), pp.2420–2428. https://doi.org/10.3892/etm.2015.2415

295. Li, Q., 2022. Effects of forest environment (Shinrin-yoku/Forest bathing) on health promotion and disease prevention - the Establishment of "Forest Medicine". *Environmental Health and Preventive Medicine*, 27, p.43. https://doi. org/10.1265/ehpm.22-00160

296. Li, Q., Morimoto, K., Kobayashi, M., et al., 2008. Visiting a Forest, but Not a City, Increases Human Natural Killer Activity and Expression of Anti-Cancer Proteins. *International Journal of Immunopathology and Pharmacology*, 21(1), pp.117–127. https://doi.org/10.1177/039463200802100113

297. Bushdid, C., Magnasco, M.O., Vosshall, L.B., et al., 2014. Humans can discriminate more than 1 trillion olfactory stimuli. *Science*, 343(6177), pp.1370–1372. https://doi.org/10.1126/science.1249168; Hansen, M.M., Jones, R. and Tocchini, K., 2017. Shinrin-Yoku (Forest Bathing) and Nature therapy: a state-of-the-art review. *International Journal of Environmental Research and Public Health*, 14(8), p.851. https://doi.org/10.3390/ijerph14080851

298. Stigler, S.M., 1978. Some forgotten work on memory. *Journal of Experimental Psychology: Human Learning and Memory*, 4(1), pp.1–4. https://doi.org/10.1037/0278-7393.4.1.1

299. Jones, T. and Oberauer, K., 2013. Serial-position effects for items and relations in short-term memory. *Memory*, 21(3), pp.347–365. https://doi.org/10.1080/09658211.2012.726629

300. Richardson, M. and Butler, C., 2023. The silk mill vision of a nature connected society. Available at: https://findingnatureblog.files.wordpress.com/2023/09/silk-mill-vision.pdf

301. Richardson, M., Dobson, J., Abson, D.J., et al., 2020. Applying the pathways to nature connectedness at a societal scale: a leverage points perspective. *Ecosystems and People*, 16(1), pp.387–401. https://doi.org/10.1080/26395916.2020.1844296

302. Nisbet, E.K. and Zelenski, J.M., 2013. The NR-6: a new brief measure of nature relatedness. *Frontiers in Psychology*, 4, p.813. https://doi.org/10.3389/fpsyg.2013.00813

303. State of Nature Partnership, 2023. State of Nature report. Available at: https://www.stateofnature.org.uk; Field, R.H., Hill, R.K., Carroll, M.J., et al., 2016. Making explicit agricultural ecosystem service trade-offs: a case study of an English lowland arable farm. *International Journal of Agricultural Sustainability*, 14(3), pp.249–268. https://doi.org/10.1080/14735903.2015.1102500

304. Richardson, M., Passmore, H.-A., Barbett, L., et al., 2020. The green care code: how nature connectedness and simple activities help explain pro-nature conservation behaviours. *People and Nature*, 2(3), pp.821–839. https://doi.org/10.1002/pan3.10117

305. Mead, J., Gibbs, K., Fisher, Z., et al., 2023. What's next for wellbeing science? Moving from the Anthropocene to the Symbiocene. *Frontiers in Psychology*, 14, p.1087078. https://doi.org/10.3389/fpsyg.2023.1087078

306. Kovaleski, C., Ahmed, N., Huckle, C., et al., 2025. A systematic review of Nature-based interventions for neurological disorders. *medRxiv*. https://doi.org/10.1101/2025.06.21.25330058

307. Prohaska, S. and Matthias, K., 2023. Effectiveness of mindfulness-based stress reduction as a nondrug preventive intervention in patients with migraine: a systematic review with meta-analyses. *Complementary Medicine Research*, 30(6), pp.525–534. https://doi.org/10.1159/000534653; Simpson, R., Simpson, S., Ramparsad, N., et al., 2019. Mindfulness-based interventions for mental well-being among people with multiple sclerosis: a systematic review and meta-analysis of randomised controlled trials. *Journal of Neurology, Neurosurgery & Psychiatry*, 90(9), pp.1051–1058. https://doi.org/10.1136/jnnp-2018-320165; Lin, H.W., Tam, K.W. and Kuan, Y.C., 2023. Mindfulness or meditation therapy for Parkinson's disease: a systematic review and meta-analysis of randomized controlled trials. *European Journal of Neurology*, 30(8), pp.2250–2260. https://doi.org/10.1111/ene.15839

308. Aitken, W.W., Lombard, J., Wang, K., et al., 2021. Relationship of neighborhood greenness to Alzheimer's disease and non-Alzheimer's dementia among 249,405 U.S. medicare beneficiaries. *Journal of Alzheimer's Disease*, 81(2), pp.597–606. https://doi.org/10.3233/JAD-201179

309. Rhew, I.C., Vander Stoep, A., Kearney, A., et al., 2011. Validation of the normalized difference vegetation index as a measure of neighborhood greenness. *Annals of Epidemiology*, 21(12), pp.946–952. https://doi.org/10.1016/j.annepidem.2011.09.001

310. Aitken, W.W., Lombard, J., Wang, K., et al., 2021. Relationship of neighborhood greenness to Alzheimer's disease and non-Alzheimer's dementia. https://doi.org/10.3233/JAD-201179

311. Liu, X.X., Ma, X.L., Huang, W.Z., et al., 2022. Green space and cardiovascular disease: a systematic review with meta-analysis. *Environmental Pollution*, 301, p.118990. https://doi.org/10.1016/j.envpol.2022.118990; Liu, C., Zhang, P., Tian, M., et al., 2023. Association of greenness with the disease burden of lower respiratory infections and mediation effects of air pollution and heat: a global ecological study. *Environmental Science and Pollution Research*, 30(40), pp.91971–91983. https://doi.org/10.1007/s11356-023-28816-y; Rojas-Rueda, D., Nieuwenhuijsen, M.J., Gascon, M., et al., 2019. Green spaces and mortality: a systematic review and meta-analysis of cohort studies. *Lancet Planetary Health*, 3(11), pp.e469-e477. https://doi.org/10.1016/S2542-5196(19)30215-3; Thygesen, M., Engemann, K., Holst, G.J., et al., 2020. The association between residential green space in childhood and development of attention deficit hyperactivity disorder: a population-based cohort study. *Environmental Health Perspectives*, 128(12), p.127011. https://doi.org/10.1289/EHP6729; Donovan, G.H., Michael, Y.L., Gatziolis, D., et al., 2019. Association between exposure to the natural environment, rurality, and attention-deficit hyperactivity disorder in children in New Zealand: a linkage study. *Lancet Planetary Health*, 3(5), pp.e226-e234. https://doi.org/10.1016/S2542-5196(19)30070-1; Louv, R., 2005. *Last Child in the Woods: Saving Our Children from Nature-deficit Disorder*. New York: Algonquin Books.

312. Ulrich, R.S., 1984. View through a window may influence recovery from surgery. *Science*, 224(4647), pp.420–421. https://doi.org/10.1126/science.6143402

313. Halawa, F., Madathil, S.C., Gittler, A., et al., 2020. Advancing evidence-based healthcare facility design: a systematic literature review. *Health Care Management Science*, 23(3), pp.453–480. https://doi.org/10.1007/s10729-020-09506-4; Bernhardt, J., Lipson-Smith, R., Davis, A., et al., 2022. Why hospital design matters: a narrative review of built environments research relevant to stroke care. *International Journal of Stroke*, 17(4), pp.370–377. https://doi.org/10.1177/17474930211042485

314. Park, P.D.N., 2023. Novel 'prescription for nature' launches in Derbyshire. Available at: https://www.peakdistrict.gov.uk/learning-about/news/archive/2023-press-releases/2023-news/novel-prescription-for-nature-launches-in-derbyshire

315. RSPB Scotland, 2022. Nature Prescriptions: supporting the health of people and nature. Available at: https://base-prod.rspb-prod.magnolia-platform.com/dam/jcr:1eb5f06d-adb0-4517-8168-70a08fcf4d98/Edinburgh-pilot-final-report.pdf

316. Swinburn, B.A., Walter, L.G., Arroll, B., et al., 1998. The green prescription study: a randomized controlled trial of written exercise advice provided by general practitioners. *American Journal of Public Health*, 88(2), pp.288–291. https://doi.org/10.2105/AJPH.88.2.288; Adewuyi, F.A., Knobel, P., Gogna, P., et al., 2023. Health effects of green prescription: a systematic review of randomized controlled trials. *Environmental Research*, 236, p.116844. https://doi.org/10.1016/j.envres.2023.116844; Robinson, J.M., Jorgensen, A., Cameron, R., et al., 2020. Let Nature be thy medicine: a socioecological exploration of green prescribing in the UK. *International Journal of Environmental Research and Public Health*, 17(10), p.3460. https://doi.org/10.3390/ijerph17103460

317. Baxter, D. and Pelletier, L.G., 2019. Is Nature relatedness a basic human psychological need? A critical examination of the extant literature. *Canadian Psychology/Psychologie canadienne*, 60(1), pp.21–34. https://doi.org/10.1037/cap0000145

318. Holt-Lunstad, J., Smith, T.B., Baker, M., et al., 2015. Loneliness and social isolation as risk factors for mortality: a meta-analytic review. *Perspectives on Psychological Science*, 10(2), pp.227–237. https://doi.org/10.1177/1745691614568352

Chapter 8

319. Carroll, L., 1895. What the tortoise said to Achilles. *Mind*, IV(14), pp.278–280. https://doi.org/10.1093/mind/IV.14.278
320. Romanello, M., Napoli, C.D., Green, C., et al., 2023. The 2023 report of the Lancet Countdown on health and climate change: the imperative for a health-centred response in a world facing irreversible harms. *The Lancet*, 402(10419), pp.2346–2394. https://doi.org/10.1016/S0140-6736(23)01859-7
321. Peechakara, B.V., Sina, R.E. and Gupta, M., 2024. Vitamin B2 (Riboflavin). *StatPearls*. Treasure Island, FL: StatPearls Publishing LLC.
322. Mahabadi, N., Bhusal, A. and Banks, S.W., 2024. Riboflavin Deficiency. *StatPearls*. Treasure Island, FL: StatPearls Publishing LLC.
323. Bashford, J.A., Chowdhury, F.A. and Shaw, C.E., 2017. Remarkable motor recovery after riboflavin therapy in adult-onset Brown-Vialetto-Van Laere syndrome. *Practical Neurology*, 17(1), pp.53–56. https://doi.org/10.1136/practneurol-2016-001488
324. Cali, E., Dominik, N., Manole, A., et al., 2021. Riboflavin transporter deficiency. In: Adam, M.P., Feldman, J., Mirzaa, G.M., et al., eds. *GeneReviews(®)*. Seattle, WA: University of Washington. Available at: https://www.ncbi.nlm.nih.gov/books/NBK299312
325. Daniel, H., 2020. *Liber Uricrisiarum*: a reading edition. In: Harvey, E.R., Tavormina, M.T., Star, S., et al., eds. Toronto: University of Toronto Press.
326. Health Care Without Harm, 2019. Health care's climate footprint. Available at: https://global.noharm.org/sites/default/files/documents-files/5961/HealthCaresClimateFootprint_092319.pdf
327. NHS England, 2022. Delivering a 'Net-Zero' National Health Service. Available at: https://www.england.nhs.uk/greenernhs/publication/delivering-a-net-zero-national-health-service/
328. Varkey, B., 2021. Principles of clinical ethics and their application to practice. *Medical Principles and Practice*, 30(1), pp.17–28. https://doi.org/10.1159/000509119
329. UK Health Alliance on Climate Change, 2023. GMC Good Medical Practice review. Available at: https://ukhealthalliance.org/sustainable-healthcare/gmc-good-medical-practice-review/
330. General Medical Council, 2024. Good Medical Practice. Available at: https://www.gmc-uk.org/professional-standards/the-professional-standards/good-medical-practice
331. Mulcahy, E. and Smith, R., 2023. Good medical practice: a missed opportunity to embed sustainability in ethical standards. *BMJ*, 382, p.1956. https://doi.org/10.1136/bmj.p1956
332. World Meteorological Organization, 2023. Provisional state of the global climate. Available at: https://wmo.int/sites/default/files/2023-11/WMO%20Provisional%20State%20of%20the%20Global%20Climate%202023.pdf
333. The Cenozoic CO2 Proxy Integration Project Consortium, Hönisch, B., Royer, D.L., et al., 2023. Toward a Cenozoic history of atmospheric CO2. *Science*, 382(6675), p.eadi5177. https://doi.org/10.1126/science.adi5177

Chapter 9

334. Ritchie, H. and Rose, M., 2020. CO2 emissions. *Our World in Data*. Available at: https://ourworldindata.org/co2-emissions
335. Capstick, S., Creutzig, F., Culhane, T., et al., 2020. Discourses of climate delay. *Global Sustainability*, 3, e17. https://doi.org/10.1017/sus.2020.13;

Islington Climate Centre, 2023. Climate delay. Available at: https://www.islingtonclimatecentre.co.uk/post/climate-delay#:~:text=%27Discourses%20of%20Climate%20Delay%27%20is,taking%20action%20on%20Climate%20Change

336. Begley, J., 2021. John Hill (1714?-1775) on 'Plant Sleep': experimental physiology and the limits of comparative analysis. *Annals of Science*, 78(1), pp.41–63. https://doi.org/10.1080/00033790.2020.1813807

337. Rattenborg, N.C. and Ungurean, G., 2023. The evolution and diversification of sleep. *Trends in Ecology & Evolution*, 38(2), pp.156–170. https://doi.org/10.1016/j.tree.2022.10.004

338. Brinkman, J.E., Reddy, V. and Sharma, S., 2024. Physiology of Sleep. *StatPearls*. Treasure Island, FL: StatPearls Publishing LLC.

339. Patel, A.K., Reddy, V., Shumway, K.R., et al., 2024. Physiology, Sleep Stages. *StatPearls*. Treasure Island, FL: StatPearls Publishing LLC.

340. Cappuccio, F.P., D'Elia, L., Strazzullo, P., et al., 2010. Sleep duration and all-cause mortality: a systematic review and meta-analysis of prospective studies. *Sleep*, 33(5), pp.585–592. https://doi.org/10.1093/sleep/33.5.585

341. Reddy, S., Reddy, V. and Sharma, S., 2024. Physiology, Circadian Rhythm. *StatPearls*. Treasure Island, FL: StatPearls Publishing LLC.; Czeisler, C.A., Duffy, J.F., Shanahan, T.L., et al., 1999. Stability, precision, and near-24-hour period of the human circadian pacemaker. *Science*, 284(5423), pp.2177–2181. https://doi.org/10.1126/science.284.5423.2177

342. Velazquez-Arcelay, K., Colbran, L.L., McArthur, E., et al., 2023. Archaic introgression shaped human circadian traits. *Genome Biology and Evolution*, 15(12), evad203. https://doi.org/10.1093/gbe/evad203

343. Knutson, K.L. and von Schantz, M., 2018. Associations between chronotype, morbidity and mortality in the UK Biobank cohort. *Chronobiology International*, 35(8), pp.1045–1053. https://doi.org/10.1080/07420528.2018.1454458

344. British Trust for Ornithology, n.d. Nightingale bird facts. Available at: https://www.bto.org/understanding-birds/birdfacts/nightingale

345. Lee, S., 2021. *The Nightingale: 'The Nature Book of the Year'*. New York: Random House.

346. Switch It Green, n.d. Website. Available at: https://www.switchit.green

347. Hayhoe, K., 2021. *Saving Us: A Climate Scientist's Case for Hope and Healing in a Divided World*. New York: Atria/One Signal Publishers.

348. McGuire, B., 2021. *Hothouse Earth: An Inhabitant's Guide*. London: Penguin.

349. Earth Overshoot Day, 2025. Available at: https://www.overshootday.org

350. Dasgupta, P., 2021. *The Economics of Biodiversity: The Dasgupta Review*. London: HM Treasury. Available at: https://www.gov.uk/government/publications/final-report-the-economics-of-biodiversity-the-dasgupta-review

351. B Lab Global, n.d. Website. Available at: https://www.bcorporation.net/en-us/; Rainforest Alliance, n.d. Website. Available at: https://www.rainforest-alliance.org

352. Soga, M. and Gaston, K.J., 2018. Shifting baseline syndrome: causes, consequences, and implications. *Frontiers in Ecology and the Environment*, 16(4), pp.222–230. https://doi.org/10.1002/fee.1794

353. Kasner, E. and Newman, J.R., 1949. *Mathematics and the Imagination*. London: G. Bell & Sons.

354. Asensi, V. and Asensi, J.M., 2013. Euler's right eye: the dark side of a bright scientist. *Clinical Infectious Diseases*, 57(1), pp.158–159. https://doi.org/10.1093/cid/cit170; Fuss, N., 1783. Eulogy of Leonhard Euler. Available at: https://web.archive.org/web/20181226041204/http:/www-history.mcs.st-and.ac.uk/~history/Extras/Euler_Fuss_Eulogy.html; Castelvecchi, D., 2015. History:

a mathematical revolutionary. *Nature*, 528(7581), pp.190–191. https://doi.org/10.1038/528190a

355. Tucker, V.A., 2000. The deep fovea, sideways vision and spiral flight paths in raptors. *Journal of Experimental Biology*, 203(24), pp.3745–3754. https://doi.org/10.1242/jeb.203.24.3745

Epilogue

356. King, D.I., Jeffery, M. and Bailey, B.A., 2021. Generating indicator species for bird monitoring within the humid forests of northeast Central America. *Environmental Monitoring and Assessment*, 193(7), 413. https://doi.org/10.1007/s10661-021-09172-1; Crozier, L., 2003. Winter warming facilitates range expansion: cold tolerance of the butterfly *Atalopedes campestris*. *Oecologia*, 135(4), pp.648–656. https://doi.org/10.1007/s00442-003-1219-2; Wilson, R.J., Gutiérrez, D., Gutiérrez, J., et al., 2007. An elevational shift in butterfly species richness and composition accompanying recent climate change. *Global Change Biology*, 13(9), pp.1873–1887. https://doi.org/10.1111/j.1365-2486.2007.01418.x

357. Biological Recording Centre and UK Centre for Ecology & Hydrology, 2025. iRecord website. Available at: https://irecord.org.uk/

358. Hill, G.M., Kawahara, A.Y., Daniels, J.C., et al., 2021. Climate change effects on animal ecology: butterflies and moths as a case study. *Biological Reviews of the Cambridge Philosophical Society*, 96(5), pp.2113–2126. https://doi.org/10.1111/brv.12746

Index

www.ingramcontent.com/pod-product-compliance
Lightning Source LLC
Chambersburg PA
CBHW050213270326
41914CB00003BA/394